The information revolution and Ireland

The information revolution and Ireland

Prospects and challenges

Lee Komito

University College Dublin Press
Preas Choláiste Ollscoile
Bhaile Átha Cliath

First published 2004
by University College Dublin Press
Newman House
86 St Stephen's Green
Dublin 2
Ireland

www.ucdpress.ie

ISBN 1 904558 07 0

CIP data available
from the British Library

Typeset in Ireland in Adobe Garamond and Trade Gothic
by Elaine Shiels, Bantry, County Cork
Text design by Lyn Davies
Printed in England on acid-free paper by Creative Print & Design

Contents

Preface

There has been an information revolution and we are either living in an information society or are about to enter an information society. At least, so proclaim newspaper and magazine articles, as well as television and radio programmes. Popular books describe the 'death of distance' (Cairncross 1997) as well as the 'third wave' which is coming after the agricultural and industrial 'waves' (Toffler 1980), and newspapers and magazines are zealous in their discussion of new gadgets and the transformation (sometimes good and sometimes bad) that these technological marvels herald. Academic writers are less certain, with some arguing that current technologies are leading to economic and social transformation (Castells 1996; Poster 1990) while others (Schiller 1985; Wood 1997) argue that the information revolution is just the Industrial Revolution with a few new frills.[1] Some have argued that new technologies will lead to freedom and empowerment (Bell 1973), while others have drawn attention to these technologies increasing the power of states or multinational corporations, at the expense of individuals (Lyon 2002; 1994; Lyon and Zureik 1996). This book is not intended to be an exhaustive or definitive discussion of the digital revolution or the information society,[2] nor is it intended to proclaim or denounce the new information society. However, whether there is a new economic, political, and social order emerging, or not; whether the new order is beneficial or detrimental to citizens; all agree that significant changes are taking place. Often, however, it is as though we are all bystanders, watching change taking place, with very little public participation in the process. The central issue in this book is that technology, including the new information and communications technologies linked with the information society, is not a force external to society and beyond the control of society; technology is an integral part of society and is acted upon and altered by social forces (Winner 1977; Williams 1974).

This book is not an attempt to predict the future; the future will result from interactions between technologies, individuals, social forces and political policies, and will differ from society to society. There has never been a single or inevitable result of technological change, and, in the context of recent technological changes, there is no 'one size fits all' outcome in which societies throughout the world will either become carbon copies of each other or mini-versions of the United States. The societies that will emerge in the twenty-first century depend on policy choices that individuals and governments make, whether by acts of commission or omission. The future is not inevitable; it is flexible and can be altered. The aim of this book is to encourage individuals to contribute to such policy choices, so that the society that emerges is one that citizens desire rather than one neither of their making or choosing. Its aim is to encourage discussion and thought rather than proclaim conclusions.

Writing a book is a solitary activity, with a lot of time spent sitting in front of a computer screen. However, it is also an intellectual statement that has been shaped by interactions, conversations, and observations shared amongst a community of friends and colleagues. I am grateful to Mary Burke, Michael Casey, Ian Cornelius, Maeve Conrick, Vera Regan and Daniel Regan-Komito for comments or observations they made on various parts of this book or discussions on issues that were raised in it. I must also include discussions with my former colleagues in the Information Technology in Society programme at Manchester Metropolitan University, John Cawood, Liz Marr and Dave Randall, who provided a stimulating environment during which I began exploring these issues. I would like to thank particularly James Wickham for comments he made on an earlier version of this book. My thanks also to the undergraduate and postgraduate students in Information Studies at University College Dublin, whose reactions to the ideas presented in lectures helped me to clarify my thoughts. The deficiencies that remain are, of course, entirely my own. My sincere thanks to Barbara Mennell of UCD Press who was always supportive through the process of writing this book. I am grateful for the support of University College Dublin through the award of a President's Research Fellowship.

LEE KOMITO

Dublin, February 2004

Abbreviations

BFI	British Film Institute
CD	Compact Disc
DVD	Digital Versatile Disc
FOI	Freedom of Information
ICT	Information and Communications Technology
IP	Internet Protocol
ISP	Internet Service Provider
IT	Information technology
NASA	National Aeronautics and Space Administration
NGO	Non-governmental Organisation
PC	Personal Computer
PIN	Personal Identification Number
RTÉ	Radio Telefís Éireann
SMS	Short Message Service
TCP	Transmission Control Protocol
VCR	Video Cassette Recorder
VHS	Video Home System
WWW	World Wide Web

Chapter 1

Introduction

It is accepted wisdom today that computer technology has been causing significant economic, social and cultural changes over the past few decades, and that these changes will continue into the future. It seems so obvious that one need hardly discuss the issue. One can 'surf the Net', getting information from anywhere in the world in a moment – pictures from space that are stored on NASA computers, or course descriptions from universities halfway across the world. Many radio stations broadcast on cable, satellite and the World Wide Web, so one can listen to a radio station from anywhere in the world. People see live pictures, on either television or computer, of events happening in distant locations. They can send electronic mail, for virtually no cost, to people halfway around the world, to one person or to a multitude of people. Digital pictures of a newborn baby can be made available, through email or the World Wide Web, to an extended family dispersed across the world within minutes. Virtual reality technology enables people to walk through a building that has not been built, and suggest alterations to the design. There are electronic shopping malls, virtual cities, and electronic banking; books or flowers can be ordered electronically, to be delivered to home or as a gift somewhere else. An article can be made available to scientists all over the world within minutes, at little if any cost to either the writer (information producer) or the audience (information consumer). Small wonder that in newspapers and television, one hears talk of the information society and the information revolution. Changes are taking place in all aspects of life, and computer technology seems to be responsible for global changes in economies and societies.

This new information society has resulted from a technological revolution. Just as there was an agricultural revolution and an industrial revolution, there is now an information revolution, the result of advances in computer and telecommunications technology. This revolution is having as great an impact on contemporary society as the agricultural and industrial revolutions did on earlier societies.[1] This technological revolution is also an economic revolution; we all participate in a global information economy, with new forms of work emerging and knowledge becoming the source of economic prosperity. It also involves social change: changes in the way we live, the way we work, how we entertain ourselves, and changes in the nature of community and society. There are new opportunities, with the possibility of greater democratic participation

for citizens, new communities for individuals, and greater access to information for people on the margins of society. But the changes are not all positive. There are new risks as well: new dangers to children, new types of crime, loss of privacy for citizens, isolation for individuals, and a permanent underclass of people who are unable to find employment in this new information economy and who cannot afford access to the increased amount of information that is now available.

Most discussions about this new information society have some common features. They are based on the assumption that new technology is transforming all of our lives, whether for good or ill, and that this transformation is inevitable. There is little we can do about it, and we had better join the information revolution before we lose our chance to benefit from it. But is a 'revolution' taking place and are we entering a new information society? While some argue that this new information society is replacing the older industrial society, others suggest we are seeing the 'informatisation of life' rather than the transformation of the old economic order (Webster 1995; 2002). Perhaps the term 'information society' is incorrect and misleading because it suggests discontinuities with the past. The suggestion of an abrupt break inhibits analysis of trends and patterns because it assumes that the past cannot be used as a guide for the future. There have been many technological innovations over the past two centuries; why should these recent developments in information technology create a new era, when the telephone or the automobile did not? There is a certain mystification in talking about an 'information revolution' and 'information societies' – a revolution implies change on such a magnitude that nothing will be as it once was. It suggests that, in the face of such a change, we cannot know what the future will hold, and we cannot control the future that will emerge.

Is a new era dawning? Is the future really so much out of our control? It is for this reason that this book has been written. It does not really matter whether there is a new information society or an informatisation of existing relationships. In either event, the nature of human beings has not changed. We are certainly living in a time of change, but technological changes, however significant or even revolutionary, are not acts of nature which alter the landscape no matter what we do or desire. The outcomes of current technological changes are not a forgone conclusion. One of the major lessons from social studies of technology is that technological changes do not have inevitable or predictable social impacts. Technological change is not something 'out there'; technology is just as much part of, and subject to, social forces as any other consequence of human action (MacKenzie and Wajcman 1985; Bijker, Hughes et al. 1987; Robins and Webster 1999). The same technologies can have different consequences in differing circumstances, depending on social and cultural factors, and the development of technology is itself the result of social forces.

Sometimes it is assumed that all societies will become mirror images of the United States; as a society approaches the US's level of technological development will it also develop similar economic, social and cultural structures? Technologies will develop differently in each society, and even the same technologies will have different consequences in different societies. Just because the United States has a strong market economy and little state support for social welfare, it does not mean that technological development and a market economy must be linked to minimal state intervention everywhere. Finland has a similar level of technological development as the United States and an equally successful economy. However, unlike the United States, Finland has a very high level of state intervention with state support for individuals and communities (Castells and Himanen 2002).The lesson for the 'information revolution' is that there is no single 'information society' emerging; each society develops its own version of an 'information society'.

This is the central rationale for this book: what society will develop as the information revolution spreads and what is the role of the individual citizen in this transformation? Whether new technologies are driving change, revolution or transformation, there often seems to be little room for individual participation in the process. We think of ourselves as powerless in the face of a vast change, just as we would be powerless in the face of a tornado or earthquake. It often seems that the best we can do is to learn as much as we can about this information revolution, so we can either ride the tiger or at least avoid being eaten by it. This view that we are powerless is a dangerous fallacy. The most important lesson to be learned is that technological change is directed and altered by the individual actions of human beings and the collective policies of governments. Both individually, and collectively as a society, we have choices – what we, as individuals, use technologies to achieve, as well as what we, as a society, devise as policies regarding the place of technology.

We are not the powerless victims of some vast and uncontrollable technological revolution; we can influence how new technologies will structure our society and we can change the technologies so as to create the society we want. As individuals, we choose whether to use an automated teller machine or a human teller, whether to inform ourselves about politics by checking websites or simply vote on the basis of information provided by politicians, whether we work at an office or at home via the Internet. But these individual choices are also constrained by social policy decisions. Should banks be required to have human tellers? Should there be sufficient infrastructure investment to enable working from home? Should governments ensure that everyone has inexpensive access so that anyone can afford to be informed? The future shape of government and society is not predetermined, just because new technologies are emerging. The future, whether positive or negative, is not inevitable. However, to exercise a vital control over our future, we first need to understand what changes are taking place, so that we can then decide, as a society, what future we want. In that context, it would then be possible to use technological developments, combined with appropriate legal and social structures, to move towards that future.

1.1 A revolution?

It may seem obvious that we are now living in an information society, but it can be quite difficult to obtain detailed evidence of how this information society or knowledge economy differs from previous societies or economies. In fact, there are many who argue that what we see around us is not actually an information society at all, but simply an advanced industrial society (Robins and Webster 1999; Schiller 1981; Stehr 1994; Webster 2002). This is not an argument that technology is unimportant, nor is it an argument that information is unimportant; it is an argument that these changes do not mark a rupture in social processes. This is not the first time that computers have been cast as agents in a revolution. In the 1990s, newspapers, television, magazines and radio all proclaimed a revolution that necessitated moving fast or being left behind, but the same revolution was also heralded in the 1970s, with the emergence of the microcomputer (Bannon, Barry et al. 1982; Bannon 1997; Robins and Webster 1999). The rhetoric was much the same. Indeed, some commentators discussed the telegraph or the telephone using much the same imagery: the death of distance, the linking of the world together (Standage 1998). So is there really a revolution taking place, or is it a creation of the overactive imaginations of overgrown boys who love computer toys? If there is a revolution taking place, why is it happening now, and not twenty years ago, or sixty years ago, or when other new communications and information processing technologies were developed? Maybe it has been going on, steadily and almost imperceptibly, since after the Second World War, or even since the development of the telegraph and telephone. In which case, maybe it is not a revolution at all, but just a continuation of the industrial revolution that first started two hundred years ago.

When we think of the agricultural revolution, or the industrial revolution, we often focus on changes in technology. In the first case, the technology of obtaining food and, in the second case, the technology of manufacturing goods. These technological developments are described as revolutions not because of the technologies themselves, but because of the far-reaching economic, political, social and cultural changes that are linked with these technological developments. The technological changes are emblematic for all the other changes in the rest of society, but it is these other changes that make a revolution, not the technology alone. How can we be certain that new computer technologies will similarly lead to such widespread economic, social, cultural and political changes as to warrant the term 'revolution'? Why is this technological change going to have more impact than the other technological changes that have been taking place for centuries, such as the telegraph, telephone and automobile?

We all know that the information revolution is linked to faster, smaller, and cheaper computers, as well as improved telecommunications. These technological changes have also led to economic change. Just as the industrial revolution meant changes in the way in which goods were produced (mass production, factories, increased importance of capital as an economic resource, in addition to labour and land), so the information revolution has led to changes in how people work, where they work, how and where goods are produced, and the emergence of new products and services. In an information society, information, rather than labour, land or capital, is the basis of wealth and power (Bell 1973). Equally profound have been the social changes, with inexpensive and rapid communication enabling people to keep in contact with each other anywhere in the industrialised world (and a large part of the industrialising world), as well as obtaining news reports from all over the world within minutes. Banking, shopping, entertainment and education can all be done at home, while close kin and friends who now live and work abroad can access the latest news from home. These all reflect changes in the way information is created, distributed and accessed. With digital information, any information – including music, video and text – can be created and recorded, then transmitted and finally received, inexpensively and rapidly, enabling anyone to have access to a vast range of information at little cost.

In the midst of all this change, how do we know when a society becomes an information society? The common measures for an information society are either technology, economy, or information based, derived from increased numbers of computers, increased numbers of people employed in 'information' professions, or increased amount of information in society (Duff 2000). But these measures are difficult to apply, and often not very illuminating. Do we measure the number of computers in a society, or do we determine what the computers are used for? Do we measure the number of people online, or what they do when they are online? If the United States has twice the computers per population than Ireland, then is it twice as much an information society? Or four times as much, if four times as many individuals use the Internet? Does a society become an information society when a certain percentage of economic transactions are electronic? Does Ireland become an information society when a certain percentage of its economic productivity derives from 'knowledge' rather than land, labour or capital? If so, how does one define and then measure the production of 'knowledge'? Is knowing the best time to harvest a crop 'knowledge'? Or knowing how to fix a bicycle? How does one measure the amount of knowledge needed to become a knowledge economy? Are people manufacturing circuit boards for computers participating in a knowledge economy or an industrial economy? If an information society is based on information, how much more information does a society need to be an information society? Does one count the number of books being published, or the amount of useless and irrelevant information that an Internet search reveals on the World Wide Web? None of the measures is clear or unambiguous, either in the way they would be applied or the meaning they would have. This is hardly surprising, because there is very little clarity about what an 'information society' is.

1.2 Assumptions

These difficulties in defining and measuring an 'information society' arise because there are so many unstated and unproven assumptions surrounding the concept. One of the major tasks in discussing the information society is to separate the concept from the assumptions that surround it. For instance, popular newspapers and magazines commonly assume that more national investment in information technology means more employment, and the incentive for countries such as Ireland to become an information society is to create more and better employment for its citizens: not only will there be more jobs, but the jobs will be more interesting. This is one of the many 'obvious' truths about the information society that turn out not to be so true. The spread of information technology does not necessarily create more or better jobs. Often, technology replaces labour, leading to unemployment, or the movement of employment to other locations in the world. In other cases, it 'supports' individual's labour, but in a way that reduces the amount of free choice and creativity that employees previously enjoyed. There are few people who would argue that working in a call centre, spending most of the day looking at a computer screen, dealing with people on the phone and having your phone calls timed by management, is an intellectually or financially rewarding way to earn a living. There will be broad changes in economic structure in an information society, but these changes are not ones in which individuals will necessarily benefit in the work they do, the salary they receive for that work, or the conditions in which that work is carried out.

There are also serious misconceptions about the impact of technology on society. It has often been assumed that technology is a good thing for society, and the more technology there is, the better for society. There were no problems that technology could not solve, and improvements in technology were inevitably going to lead to improvements in society at large. The fallacies inherent in this view are now clear. There are many problems which technology alone has not been able to solve and may never be able to solve, including problems of inequality, disease and malnutrition, to mention a few. Not only that, but technology can itself create new problems: nuclear waste, global warming, increased cases of cancer, antibiotic resistant super-bugs. The list could go on and on.

Before a technological 'advance' is welcomed, its impact must first be assessed. This is easier said than done, because it is actually not easy to predict in advance the impact of technological innovation. Although it is easy to forget this now, no one predicted as the Internet developed how significant electronic mail would become. It was almost an afterthought in the development of Internet protocols, and, initially, only those with computer expertise (and who were also highly motivated!) could master the intricacies of addressing and sending electronic mail. IBM is known for the mistake of not keeping control of the operating system being used for its personal computers (thus enabling the growth of Microsoft), a mistake it made only because it did not anticipate how successful PCs would become. It is also known for another enormously incorrect prediction, thought to have been made by Thomas Watson, the founder of IBM. He suggested, in 1943, that the world market had room for about five computers.[2] More recently, the popularity of the World Wide Web took people by surprise.

These surprises extend to other technologies as well. The popularity of text messaging (Short Message Service or SMS) was not predicted by mobile phone designers, with designers trying to catch up with users who were using technologies in ways that had not been anticipated. Conversely, even excellent technologies may not become popular. We all use the VHS format for videotape machines; few remember BetaMax, which provided excellent recording quality and initially dominated the market (Shapiro 1999: 94). However, it provided only hour-long tapes when people wanted to record and play two-hour movies, and so quickly disappeared.[3] Similarly, the early versions of the Apple Macintosh were often far superior to the early IBM Personal Computers, in technology and ease of use, yet Apple computers now account for about ten per cent of the microcomputer market. To put it simply, technology does not inevitably or predictably lead to specific economic or social structures: it is not deterministic. As often as not, the design and development of technology is, itself, a consequence of broader economic or social forces. Thus, developments in information technology do not have predictable or inevitable social consequences.

Yet people view many social consequences of technology as inevitable. Improved work conditions are assumed – such as working from home in rural areas, better paid and more interesting employment – and also political benefits. For instance, a popular catchphrase is that 'information is power'. The early computers were expensive and large and only wealthy corporations could afford them. The coming of the personal computer permitted access to significant computing power by private individuals who had only limited financial resources; individuals could access information and, indeed, create and distribute information as effectively as large organisations. There is an assumption that new technologies permit easier access to information, and access to more information. A web page made by an individual looks as good and may be as effective as a web page made by a multimillion Euro company, and a small company can market their products as effectively as a large company. Individuals can now access information that governments and large organisations might previously have suppressed or censored, such as information about nuclear safety.

In practice, information has to be found and evaluated before it is useful. A search on the Internet reveals a great deal of information; deciding whether the information is accurate or relevant is not so simple, nor is deciding how it should be interpreted straightforward. A perfect example is that individuals can now obtain medical information on the Internet, rather than be dependent on their own doctor. A search for a medical condition on the Internet reveals much information that is too complex and technical for most people; the sites that are helpful are ones in which someone else has simplified the material and presented it in an easy-to-understand manner. In other words, the individual is still dependent on someone else. Furthermore, the website cannot determine whether the information is relevant to the particular person; that still requires an expert who knows the person. In the end, the person remains dependent on organisations and reliant on the expertise possessed by others.

1.3 Ireland

The information transformation is global in impact, but local in manifestation. International restructuring has its impact by changing particular individuals and societies, and so one must situate discussion in the experiences of particular people and particular societies. Books on this subject often focus on the United States, partly because the authors are often located in the United States and partly because the United States is viewed as the exemplar or future model for other societies. The premise of this book is that no single model exists, and societies will accommodate to new technologies differently. Furthermore, each society's perspective and experience of the global transformation is different because each society's vantage point is different. The global village is composed of many different communities, each of which is linked to the global system in different ways. Ireland provides a useful contrast to the United States and Japan, as well as to other European countries, and is the 'vantage point' of this book.

The information society debate has a different hue in a geographically peripheral and post-colonial society such as Ireland. Ireland has been striving to find a niche for itself in a world economy. Ireland came late to industrial development, providing raw materials for industrial processing in other countries and then consuming the products of those raw materials, while other countries made the profit from the conversion of raw materials into consumer products. It is difficult for Ireland to increase its industrial base, especially since it lacks many of the resources necessary for industrial growth. The information society seems to represent a move from an industrial economy to a knowledge-based economy. New technologies offer great promise for future economic growth by enabling Ireland to become the value-added producer of information products rather than industrial products. In the 1980s, Ireland pursued a strategy of encouraging multinational corporations to invest in high technology industry, while more recently industrial policy has moved from electronic manufacturing to software (Wickham 1987; 1997; Ó Riain 1997). The hope is that Ireland will produce knowledge or provide services, rather than industrial or consumer goods, and so compete in the global economy. The raw material of knowledge production is information, and the economic profit to be made from using education, knowledge and experience to transform such raw materials into market commodities can be impressive, as evidenced by Silicon Valley in Northern California and Route 128 in Boston (two of the best-known localities in the United States for technological innovation).

The information revolution is also linked to an expectation that new technology will overcome barriers of distance, and especially the economic impact of such barriers. Ireland's participation in a global industrial economy has been hindered by high transportation costs. With new technologies permitting rapid and inexpensive information transfer, perhaps any society, including Ireland ('an island behind an island', as Ireland has been described), will be able to participate fully in the world economy on the same terms as societies which are close to the economic centre. If the barriers of isolation and distance are irrelevant in an information society, then Ireland can compete with global centres such as New York, London and San Francisco. For that matter, peripheral areas of Ireland could also maintain their economic viability. Instead of people going to where jobs are located and rural areas becoming depopulated as people move to the urban centres (whether Irish centres such as Dublin and Cork or urban centres abroad), perhaps information-based jobs will come to people in rural areas.

Encouraging economic development in Ireland is a rather complex process. In the early days, Irish governments encouraged foreign investment, providing cheap but well-educated labour, along with tax incentives, in the hope not only of creating employment, but also creating linkages that would encourage the development of indigenous companies (Wickham 1997). However, that process is not straightforward. Labour costs become an issue.

At one point, Ireland was able to compete in the world economy because the cost of producing goods was low, primarily owing to low labour costs. But selling products in an open global economy based on cheap labour provides only a short-term benefit; inevitably there will be other economies with even lower labour costs and companies may move operations to those lower cost environments, unless there are major differences in infrastructure. The goal is to create employment that is not mobile, not having to compete with low cost labour economies at one end, and high education and research economies at the other. The aim is to provide an infrastructure to support a middle ground for an economy like Ireland's.

Investment in technological infrastructures (high-speed telecommunications, for instance) will return profit, but only as part of a broader foundation for knowledge production. It is still necessary to have the people who can contribute to this knowledge production. This is achieved through education and experience, but with these two factors come other issues. Can such activity be spread throughout the country, including rural areas, or does there need to be a critical mass in one locale, to ensure innovation and growth (Ó Riain 1998; Grimes 2000; Komito 1992)? Must individuals pay for the training that enables them to sell their expertise on the open market, or will the state subsidise that training? If the state subsidises the training, should the state get a return on its investment if the person leaves Ireland to make money elsewhere? Should the companies that get the benefit of this training also contribute to the cost of training?

Cultural aspects of the information society also have a resonance for Ireland. Terms like diaspora and labour mobility have special meaning for Irish people, since Irish society has been conditioned by generations of Irish people leaving Ireland to work in the United Kingdom, the United States, Canada and Australia, to name a few. One attraction of a new information economy is to provide employment to keep a young population from emigrating. However, an educated workforce remains a mobile workforce, even if that mobility is now voluntary, and so Irish people are dispersed throughout the world. The Irish diaspora is also composed of descendants of Irish emigrants whose links with contemporary Ireland are tenuous and often based on a romantic fantasy derived from historical events that are now irrelevant. What impact will new Information and Communication Technologies (ICTs) have on this dispersed population? What impact will they have on those who grew up in Ireland and are now working abroad? When people are able to travel back to Ireland frequently because of cheap transport, are able maintain to contact with friends and family through email, and can listen to Irish radio, watch Irish TV and read Irish newspapers through the World Wide Web, what impact will ICTs have on them and their sense of national identity? If a person lives in one place, physically, and yet continues to participate in Irish society through technology, do they retain their Irish identity, or do they assimilate the cultural beliefs and social practices of the society in which they currently live?[4] How will this new Irish diaspora change the nature of Irish identity, even for those who remain in Ireland?

The political dimension of the information society also has a relevance for Ireland. New technologies offer new ways for citizens to participate in the making of political policies and new ways to access government services. It is no longer necessary to depend on politicians to represent and articulate the aggregated public opinion of a constituency; citizens can express their views directly to those making decisions. This has strong resonances in Ireland where there has been a tradition of clientelist politics – that is, depending on politicians for information about government services, or assuming that political intervention is necessary to obtain benefits to which people are entitled (Komito 1984). New technologies permit direct access to government information and services, thereby circumventing the politician's monopoly over information and access. This may undermine the traditional dependency of citizens on politicians, thus changing the nature of politics in Ireland (Komito 1997; 1999).

1.4 Outline of the book

This book examines the impact of information technology, and especially the capacity to digitise information, on society. While technology does not determine society, there is no doubt that any technology permits and even encourages certain behaviours or structures, while making other behaviours or structures more difficult or more costly. Technologies of information have always had a special impact on society, but especially in recent decades. However, these technological developments, as well as their social correlates, can be understood only in the context of the history of technologies of creating, distributing and consuming information over past centuries. Chapter 2 will discuss the history of information and technologies for manipulating information, to illustrate the economic, political, social and cultural impacts that technologies have had in the past. Chapter 3 will then explore the actual technological developments that have led to a move to digital (and not just electronic) information. Most of the discussion of technology focuses on its economic impact: the development of a post-industrial society. This link between technology and economic life will be explored; focusing particularly on whether this new 'information society' can adequately be defined or understood in economic terms in chapter 4. It will then be suggested in chapter 5 that an economic definition of an information society is inadequate, and a broader social definition will be proposed. The political implications of these technologies will be explored, in terms of participation, administration, and regulation, in chapters 6 and 7. Chapter 8 will discuss broad changes in the structure of societies, and, following this, the impact of these technologies on individuals, societies and communities will be discussed in chapters 9 and 10. The final chapter will review the role that individuals can have in influencing the impact of these new technologies.

Chapter 2

History of information

People often think of the term 'information technology' as a single thing, rather than as the combination of two terms: information and technology. The first step is to distinguish between the two. Technology can be variously defined; for instance, the popular Microsoft Encarta Encyclopaedia defines technology as 'processes by which human beings fashion tools and machines to increase their control and understanding of the material environment' (Merritt 1995), while the Oxford English Dictionary defines it as 'the application of scientific knowledge for practical purposes'. The term is derived from the Greek words *tekhnē*, which refers to an art or craft, and *logia*, meaning an area of study; thus, technology means, literally, the study, or science, of crafting (Oxford English Dictionary 1998). When combined with another term, technology refers to the set of practices or techniques associated with the application of knowledge to achieve a particular goal, often by manipulating objects. Thus we talk about automobile technology, meaning the practices, techniques, knowledge, and often tools, associated with making, maintaining and using cars, or similarly phone technology to discuss techniques, knowledge and tools associated with making, maintaining and using telephones. In this perspective, making spears or knives is a technological process, just as bronze, iron and pottery are all products of technological processes. As a species, we are dependent on our ability to influence and alter our external environment through tool use. We use tools to alter the external environment and to create a 'buffer' between ourselves and that environment (e.g. building houses, making clothes). So important has this been for the evolution of our species that one of the earliest and most enduring of distinctions between *Homo sapiens* and other animals has been to describe human beings as 'tool makers' (e.g. Oakley 1957; but see Pfaffenberger 1992; as well as Ingold 1987; 1988; 1993).

Since technology is a set of techniques surrounding a particular goal or object, then any set of techniques or technologies used to manipulate information is an 'information technology' and so includes any information manipulation which is not intrinsic to the human being. Speech conveys information, but is not an information technology as there is no manipulation of objects or use of practices extrinsic to the human being: human beings need only what they are born with in order to speak and understand the speech of others. On the other hand, the printing press is a technology to manipulate information (and thus an 'information technology'), and writing is an information technology, since both depend on the extrinsic objects (printing press, paper and pen, or stone tablets and chisel). Human beings were using technology to manipulate information long before computers were developed, and these technologies have been integral to the development of society for nearly ten thousand years.

Arguably, before that, information, even without technological manipulation, was critical to the success of the human species. Human dependence on information has been part of the evolution of the species: we are born with very little 'innate' knowledge and we acquire information as we grow up, and manipulate and communicate information as members of social groups.[1] If information has been central for tens of thousands of years, and information technology has been central for almost ten thousand years, then why so much fuss about computer technology in the last few decades? The only way to understand this is to review the history of information and examine the economic, social, political and cultural impact that technologies have had on information. This will enable us to see how computer technology has led to the emergence of digital information.

2.1 Information and human evolution

Debates about data, information and knowledge abound (see Buckland 1991). As a convenient set of distinctions to start with, data consist of 'raw facts', while information is 'meaningful' data, and knowledge is information integrated with other information by the human mind. Data in isolation are of little relevance to a human being or, indeed, any animal. All animals are surrounded by facts – the colour of the sky, the temperature, whether it is day or night, but only some of these data are relevant and so can be considered information. All animals need information about the external environment in order to obtain food and shelter; this information is processed, and then acted upon. If the animal is in danger from different predators at night than from predators during the day, then whether it is day or night becomes informative. Data that are informative are processed; the rest is filtered out and ignored. Human beings spend their lives ignoring sensory data – things we see, hear, or feel – because it is not relevant. Most of the time, we are not even being aware of this filtering process – we automatically and unconsciously do not notice things. Was that car red or green? Was there grass growing through a crack in the pavement? They usually just do not matter, and so we filter those data out.

How does an organism decide which data are 'informative' and which data are to be ignored? For many animals, this filter is innate. The criteria by which these animals determine what is 'information' (and thus to be taken account of) as compared with 'data' (which are ignored) are determined through genetic programming. In addition, their response to 'information' is determined in advance and is unalterable. Animals react automatically and unconsciously to threats and opportunities in the environment. However, as the environment changes, what is important and is not important changes, and so the filtering criteria must change. Since the filtering is 'built in', the change takes place through natural selection over a number of generations; it is slow and often lacking in sophistication. As long as the external change is also slow, then a slow change over time is sufficient. However, if the external change is rapid, then the slowness of change may prevent the animals, and indeed the entire species, from adapting, and they may make what are now inappropriate interpretations and responses. Individuals may die, and the entire species may become extinct. Early studies of animal behaviour demonstrated this when young birds would peck on a cardboard cut-out that duplicated the visual stimuli associated with a mother bird bringing food for its offspring. Even though no food was forthcoming, the young birds had no option but to continue to react automatically and unconsciously, exhibiting inappropriate responses to stimuli (Hinde 1974; 1982; Tinbergen 1969).

Mammals in general, and primates in particular, possess two crucial and related characteristics that enable them to avoid some of the dangers of such rigid filtering of data and rigid responses to information. Most animals have to know from birth what constitutes danger or opportunity, and how to react to it, and so must depend on in-built, or genetically determined, filters. Mammals give birth to live young that are born in a dependent state; the young require nursing in order to survive. Since young mammals are protected and fed after birth, they do not need to know at birth how to survive. Instead, they have the luxury of learning to some degree what constitutes meaningful data ('information'), and what constitutes the appropriate response to that information. They may learn from observing adults, they may be 'taught' through demonstration, or they make 'learn' through trial and error. Innate and rigid criteria for evaluating and reacting to data are less important for mammals than for other animals; for mammals, and especially primates such as human beings, it is the process of learning that is central.

In addition, some mammals are social; they exist in social groups and they co-ordinate their activities. This is especially true for primates, of which human beings are the best-known members (Hinde 1974; Crook 1972; Jolly 1972; Lancaster 1975). It makes it possible for young to learn not just from the environment, but also from other members of the same species. Information can be passed from one generation to the next within a social group. However, such co-ordination requires communication – the exchange of information among members, and so mammals have some system of communication (though often itself limited to genetically determined signals). Thus social mammals (including human beings) are groups of individuals who learn a shared set of 'filters' and learn a common means of assigning significance to the information they receive. The advantages of social learning are multiple. What has been learned once can be changed, not only from one generation to the next, but even during an individual's lifetime, as circumstances change. Furthermore, because these criteria are learned, rather than innately coded, the filtering can be far more sophisticated as well as more flexible. This makes it easier for them to adapt to changing environmental circumstances. It enables them to develop more sophisticated means of evaluating 'data' and more sophisticated responses to that information.

Considering how long the human child's dependency period lasts, one could say that human beings are born at an earlier stage of development (as compared with other species) and so are, at birth, the most dependent of any species on being cared for by others. Human beings are thus the most dependent on learning and least dependent upon inbuilt filters of any species. This flexibility enables us to adapt to our external environment through learning and altering our behaviour and our responses – as opposed to many other animals, which must change their physiology, through evolutionary adaptation, as an adjustment to changed external conditions. Our ability to develop flexible and collective responses to the external environment, and pass that knowledge on to subsequent generations, has enabled us to adapt effectively to the world.

2.2 Non-verbal communication

Crucial for any group is the facility to communicate amongst members of the group. Most animals communicate with each other using 'non-verbal' signals, such as sounds, movements or expressions (Hinde 1974). These messages carry information about emotions or states of mind, rather than information about the external world. Thus, a cry indicates 'fear', but not 'a lion is attacking'. Fellow members of the species react, based on the context of the message, in order to decide if the 'fear' message means a lion is attacking or a vulture is swooping. The difference between the two threats may be important, as the defence reaction will differ, but the difference comes from the context of the message rather than the content of the message (Smith 1977). As a communication system, non-verbal communication is quite limited. There are only a limited number of signals available – there are only so many facial expressions or cries that an animal can make. These signals are heavily dependent on context to be meaningfully interpreted. Since the signal conveys only limited information and the context may be ambiguous, this may lead to inappropriate or ineffective responses. There is also a limited amount of information that can be conveyed from one generation to the next. Non-verbal signals are also rigid in their link between signal and meaning; usually the link is not learned but is genetically encoded (see Scheflen 1974; Weitz 1979). Thus, changes can take place only slowly, through evolutionary processes.

Human beings also communicate using non-verbal signals. Non-verbal communication is often seen as more 'real' (meaning, usually, more honest) than verbal communication because non-verbal communication is thought not to be subject to conscious or intentional manipulation or control. While human beings can invent sentences to say anything they want, they unconsciously 'exude' non-verbal signals that reveal their state of mind. Since non-verbal communication is more trustworthy – especially about emotion and feeling – than words, books on body language purporting to show what people 'really' feel are always popular (for early examples, see Morris 1977; 1987). For the same reason, we can trust what people say 'face-to-face' more than over the phone because we think that posture or tone of voice is a more accurate barometer of truth than speech. Non-verbal communication differs from verbal communication in form and content; these non-verbal messages about emotional states cannot simply be 'represented' in verbal form. Although human beings try to convey, using language, the same information as is contained in non-verbal signals, it is often very difficult, if not impossible, to convey in speech or print the information conveyed through non-verbal communication.

2.3 Oral language

Non-verbal communication may be honest, but it is limited and inflexible. Language, on the other hand, is unlimited and amazingly flexible. Even though some other primates are capable of symbolic communication and perhaps language, they are not dependent on it for survival, as are human beings. The essential difference between non-verbal communication and speech is that in speech there is a symbolic and arbitrary link between utterance and meaning. Whereas the meaning of non-verbal signals is often linked to the signal itself (louder means greater urgency, for instance), the meaning of a word or sentence bears no relation to the sound of the word. Human speech uses a limited number of sounds, which, when combined, produce an unlimited number of words, which, in turn, combine to form an unlimited number of sentences, and the meanings associated with words and sentences have to be learned. Learning a language means learning a set of rules by which we can create infinite sequences of words and sentences out of a finite number of building blocks, which other speakers can accurately understand by applying similar rules. We constantly hear utterances that we have never heard before, and create utterances that others have never heard before, and yet we can decipher their meaning. This requires a sophisticated set of rules, to create as well as to understand new utterances (Crystal 1987; Hudson 2000).

This use of a limited number of sounds to create unlimited and diverse meanings makes it possible for a vast amount of information to be conveyed. There is literally no limit to what can be conveyed. Language enables individuals to describe events or actions not present (Hockett 1960). Events in the past or future or events experienced by others can be described, thus expanding the human horizon. Individuals (including children) can be told about things, rather than having to be shown things or having to experience the things themselves. They can be told what a flood is, why it is dangerous, and what to do if it happens, instead of having first to experience the flood to find out what it is and that it is dangerous. If a new danger, like fire, emerges, then the danger can be described instead of having to be experienced, with appropriate responses also described rather than being demonstrated. It is flexible, allowing the creation of new adaptations to new circumstances and allowing new knowledge to be passed from one person to another. It represents the most impressive information system ever devised.

Impressive as speech is, there are intrinsic limits to it. Until about 10,000 years ago, information could be conveyed verbally but could not be recorded in any permanent or enduring form. This restriction imposed a spatial limit on co-ordination and social action, as a limited number of people's actions could be co-ordinated by word of mouth. To spread information across a large distance takes a long time and a lot of energy, since each person has to be told and it takes ten times as much effort to tell ten people as one person. Even if one relies on a pyramid effect to reduce the transmission 'cost' (each person telling two or more others), accuracy suffers. We are all familiar with the children's game of Chinese Whispers, which illustrates how susceptible speech is to distortion. There is little that can be done to reduce this distortion; the fact that people speak with slightly different accents or different styles easily leads to misinterpretation of a spoken utterance.

There is also a limit to the amount of information that can be remembered, which limits people generationally as well as spatially. How much can be remembered and retold so that it can be passed on accurately from one generation to another? Often, succeeding generations must remake many of the same discoveries over again, and it is very difficult to build upon earlier knowledge. Oral memory is subject to alteration; it is not fixed in either time or space, and will change according to social circumstances.[2]

For these reasons, human communities dependent on oral language have tended to be relatively small in size. Hunting and gathering communities before the Agricultural Revolution often numbered no greater than 250–500 individuals. There were, of course, many, many such communities ten or twenty thousand years ago, but each tended to have its own internal structures and there was little co-ordination between them (Lee and DeVore 1968; Ingold, Riches et al. 1988; Woodburn 1982). Such communities also had limited technological sophistication. This was partly because of the difficulty of accumulating information when it had to be passed on orally from generation to generation. Thus, ten thousand years ago, the development of human societies reached a plateau, owing to the limitations of oral information.

2.4 Writing – the first information 'revolution'

Technology, as we noted on p. 13, is a set of techniques used to achieve a particular goal or to manipulate objects. Non-verbal communication and language, while important for communication of information, do not depend on technology. Writing, on the other hand, requires information to be inscribed on physical media, whether stone or paper, which stores or records information. Various 'technologies' are required in such a process: whether it is stone tools to carve on stone or paper on which to record information, technology is integral to the process. If an information technology is a set of techniques or technologies used to manipulate information, then written language represents the first human 'information technology'. This technology may not seem revolutionary, since writing seems like speech; it simply encodes speech in a permanent form. Like speech, writing is based on a limited set of symbols that are combined to create infinite numbers of words, which in turn create an infinite number of sentences, and, like speech, the letters and numbers themselves have no intrinsic link with the meanings associated with them. However, this technology permits information to be stored external to the human being. It is independent of the individual who uttered it, and, as a text, can be accessed by individuals who are distant in time or space from the individuals who first recorded the information. Speech is linear and sequential; written information can be viewed more flexibly. Individuals can look at written text and go from the middle back to the beginning, or they can jump to the end; they can even create different 'paths' by linking bits of text with lines or arrows. The impact of writing as an information system has been revolutionary.

The development of writing is linked with the agricultural revolution. This began with the domestication of plants and animals 12,000 years ago, with villages emerging from 9000 to 6500 BC. Land became an economic and political resource, to be controlled, allocated and inherited, permitting investment in fixed capital resources (including buildings). Increased efficiency in food production facilitated the development of states and stratified societies with political elites (religious figures, rulers, warriors and scholars). These changes are also linked with the development of writing. The earliest writing systems date from five thousand years ago (3000 BC). They developed over a period of a thousand years (Daniels 2003) once domestication permitted a division of labour, with farmers producing enough food to feed others who could then devote themselves to intellectual activities. It was probably also a response to problems posed by the emergence of the larger and more complex urban civilisations than the agricultural revolution could sustain. Some system of recording information was necessary in order to co-ordinate activities on an increasingly wide scale, and the earliest writing systems were used for administrative purposes: such as making lists (recording information) for collection of taxes, and keeping track of military supplies (Beniger 1986; Goody 1977; Ong 1982).

Writing was as important an information revolution as was the development of agriculture an economic revolution because it changed the way information was created, distributed, accessed and reproduced. Speech lasts only as long as the utterance itself and is distorted over time and space as it is reproduced, but recorded speech (writing) continues to exist after the event. This permits the accumulation and the diffusion of information over time and space without distortion (at least until the original written text is reproduced). With writing comes history, the possibility of a record of events not subject to inevitable alteration in the retelling. There is no history without archives, and no archives without written language. Once information can be accumulated, then the amount of information available to a group is no longer limited by the capacities of human memory. It then becomes possible to build on previous information, instead of starting from the beginning. When people need information, they can simply read existing texts, rather than 'reinventing the wheel'.

Speech and writing seem to differ only in that one is a stored version of the other, but there are subtle and important differences between them. Speech provides immediate feedback; the speaker knows the audience reaction, and alters subsequent speech accordingly. Speech content is targeted with the audience in mind, and misunderstandings are quickly corrected. The speaker and audience share a context, which helps the speaker phrase things in a way that the audience will understand. Indeed, this context is often crucial: when people in a group know each other very well, phrases or ways of speaking that make sense to people 'in the know' are meaningless to outsiders. We are all aware of slang words and specialised vocabulary known only to members of the same group, such as teenagers and doctors. Words that are meaningful for such a restricted audience are only to be used when the speaker is certain that his or her audience shares the same background.[3]

In contrast, written language is public expression to an unknown and potentially disparate audience. The readers are removed from the writer, and it is hard to predict, when writing, who the readers will be, or what their preconceptions. There is no immediate feedback by which the author can verify if the writing is being correctly interpreted. The author can therefore take little for granted in terms of shared context, and it is necessary in written language to avoid context-dependent meanings. This leads to more explicit and less ambiguous information; unlike the 'fear' cry that means different things in different contexts, written language has to be context-independent and mean the same thing regardless of the context.[4]

In speech, the person and the message are intertwined, but in written language information is distinguishable from the person conveying the information. Writing detaches the message content from the producer, extending information beyond the spatial and time confines of the people who produced it. Words become enduring objects, which can be examined, analysed, and discussed, all of which puts a distance between a person and their verbal or written actions (Goody 1977: 150). Written texts permit and encourage a scrutiny and scepticism that is not possible with oral information. While not trying to suggest that critical thought is absent in societies without written language, it is true that the nature of criticism has to be different. As Goody (1977: 37) says of written texts:

> No longer did the problem of memory storage dominate man's intellectual life; the human mind was free to study static 'text' (rather than be limited by participation in the dynamic 'utterance'), a process that enabled man to stand back from his creation and examine it in a more abstract, generalised, and 'rational' way.

It can be argued that writing encouraged the idea of 'facts', and that scientific analysis was only possible once information existed as facts that were separate from their social context. This is not to say that analysis is impossible in pre-literate societies. Indeed, scepticism is also possible in traditional societies; for instance, individuals may query witchcraft, divination, or other mystical pronouncements. However, Goody argues that the accumulation of scepticism is difficult:

> [It is] . . . difficult to develop a line of sceptical thinking about, say the nature of matter or man's relationship to God simply because a continuing critical tradition can hardly exist where sceptical thoughts are not written down, not communicated across time and space, not made available for men to contemplate in privacy as well as to hear in performance (Goody 1977: 43).

The tradition of scientific investigation can develop only once written language has developed.[5]

Writing extended, and indeed revolutionised, information production, transmission, access and reproduction; yet, any technology must involve restrictions imposed by the cost of using the technology. In the case of writing, there remains the cost of producing written text. One person must write the text, which involves a cost in both labour and materials. The same cost is incurred in reproducing the text; there are no economies of scale, since a second copy costs as much to produce as a first version. The cost of producing a text thus limits the number of texts produced. With a limited number of texts, stored in a few locations, people had to travel to gain access to them, which was an additional cost. The great library of Alexandria was founded in the middle of the third century BC and became a repository of some 400,000 manuscripts (Levinson 1997: 19), but one had to travel to the location to access it, or the single text had to be transported to the reader. Travel is not the only cost associated with access to single copies of information; there is a cost in storing text, since paper takes up storage space. Equally serious is the cost of eventually replicating the text, since manuscripts deteriorate over time and must be renewed.

There is also a cost for accessing written information. Just as one needs a computer and network connection to access information on the Internet, one needs to know how to read (and read the particular language) to access written information, and one needs to know how to write in order to add to the store of written information. One may easily learn to speak and it is almost impossible not to learn to speak. In contrast, one has to learn, and be taught, to read and write. There is thus a cost, in terms of time and labour, on the part of the person learning and the person teaching. Many people were therefore not literate, because it was not seen as worth the investment. The ability to read was also seen as too powerful to be given to all and sundry; reading and writing were elite activities, with a political dimension. Writing does not just convey thoughts, it also records data about people and events. Writing developed to control non-primary relationships beyond the confines of immediate face-to-face ties: exercising control over citizens, and helping in the administration of an increasingly complex and large-scale society. Literacy conferred power and would be at the very least a waste, and at the worst dangerous, in the hands of peasants. This lack of literacy excluded them from the control technologies of records, taxes and laws, and made them dependent on others. Written language was one of the first examples of power deriving from differential access to information.[6]

2.5 Printing press with moveable type

A significant cost limitation for written text was the cost of producing multiple copies of the same text. Producing subsequent copies required as much labour as the first. Since it was costly to produce any text, and just as costly to produce multiple copies of text, the amount of information in circulation was severely limited. A less important, but still significant, problem was that copying texts often introduced errors, as scribes made mistakes copying from one text to another. For these reasons, the printing press was an important innovation: it permitted multiple copies to be made of the same information at very little additional cost. The original printing presses depended on a large wooden block, onto which the text and/or images were carved. Once carved, it was possible to make additional copies of texts at relatively little extra cost. Though a significant amount of labour was required to produce the initial text, multiple copies, which could then be stored in multiple locations, became possible. The process of creating the carving was still expensive in time and labour, so even additional copies were still too costly for mass consumption. But it did decrease the barrier to accessing information since multiple copies also made it possible for copies to be stored in multiple locations. This reduced travel costs and increased the number of people who could access texts, even if they could not afford to possess the texts themselves.

Each block was created as a single entity, each page required a new block to be carved, and a new book required a large number of such blocks to be carved. The invention of moveable type meant that a page, previously printed from a single block, could be printed from a set of previously existing smaller units: letters. These existing units of typeface could be recycled to create a new block of type, thus reducing the cost of producing information. Creating a new page was a simple matter of rearranging existing letters to form new sentences. It made possible the rapid printing of new information, and further reduced the cost of making multiple copies of that information.

The printing press led to the first 'information revolution' – a virtual explosion in the amount of information available. With a reduction in costs of information production and reproduction, more information came into the public domain as the cost of ownership dropped. As the labour and materials cost of production dropped, it also became less crucial to decide if the information was of sufficient value to warrant the previously high cost of publication. Less significant information was printed – or, more accurately, information which was less significant by the standards of the very wealthy. For instance, the Bible had always been deemed important enough to publish, but only in its Latin form. Now it became affordable to publish translations in local vernacular forms: German, French, English. Literary as well as scientific texts could be published; texts that had not previously been deemed to be of fundamental human importance could be published. Mass public opinion, rather than elites, decided the value of the information. For example, Thomas Paine's *Common Sense* sold 120,000 copies in three months (Shapiro 1999:5; see also Eisenstein 1983; Feather 1998).

The availability of multiple copies of texts, affordable by many different people residing in different locations, has had a significant cultural consequence. We have all had the experience of losing ourselves in the imaginary world of a book, film or play. These worlds of imagination were, prior to mass publication, individual experiences of isolated readers. With multiple access to the same text, people scattered over a wide area could participate in a common symbolic experience. This has been a common explanation for the development of national identity (Anderson 1991). Printing made it possible for a wide range of people to have a common experience, which created the illusion of shared community; this imagined community is the basis for national identity. People whom you have never met – never have any prospect of meeting – are perceived as part of the same community and to be treated as one would treat any other member of a community. Reading texts (books, magazines, newspapers) at the same time is one example of shared experience, but print also permits other shared experiences. The creation of textbooks – a canon of what should be read – is possible, so that common schools and learning a common history help create a sense of common heritage. In a variety of ways, print makes a common shared experience possible. Such a common identity is crucial for nationalism – why go to war and risk death to protect someone you do not identify with and have nothing in common with?

It has also been suggested that the printing press promoted national identity in another way. For many centuries, for Catholics, the Bible was available only to a few and was always in Latin. With the printing press, came not only an increase in the number of Bibles available, but also an increase in the number of languages in which the Bible was printed. It became possible, as Martin Luther argued, for people to read the Bible for themselves, instead of depending on the Catholic Hierarchy. Thus it can be said that this facilitated the Reformation (see Levinson 1997), and also the development of national rather than universal literatures. This encouraged the segmentation of the information world – no longer was there a single universal language for intellectual activity, but a variety of different languages for different groups. While the printing press permitted the globalisation of information, since multiple copies of the same information could be distributed anywhere in the world, it was, in fact, more significant in facilitating localism. It became affordable to produce local information for local consumption in the local language.

The facility to produce multiple copies of the same information also assists the process of bureaucratisation. It is possible for administrators to consult the same set of rules and procedures even when they are scattered in different locations. This consistency of action is the hallmark of bureaucratisation, and the development of technologies to store information and distribute that information throughout an administrative system was crucial to the development of government (Weber 1978; Ellul 1964; Mumford 1934; Beniger 1986; Giddens 1985).

2.6 Ephemeral information

Books entail a cost to produce and to purchase, and so tend to contain only information intended for long-term use. Why spend money to produce or buy information unless it remains accurate for a long period of time? Since the invention of the printing press and moveable type, improvements in printing presses and the production of paper have reduced the cost of producing and purchasing print texts. These publications have a lower quality of newsprint, and are cheap to produce, since they are not intended to be stored. This is appropriate for information that is relevant for a short period only, and will be disposed of after that point. In the nineteenth century, technological improvements reduced the price of newspapers, and increased newspaper circulation resulted as everyone, not just the wealthy, began to read about current news (Curran 1977; Moore 1989). This, linked with increasing levels of literacy among the public, meant the development of a public mass media such as newspapers and magazines. With newspapers and magazines, information begins to acquire a time value – it has to be 'current' to be valuable. Such public media are even more important than books in terms of creating and sustaining a common experience and common identity, since not only is everyone 'consuming' them, but a particular issue of a newspaper or magazine is read by everyone at the same time. These common experiences are the basis for conversation and discussion, thus creating even more common experiences. Such mass media, even more than earlier books, were an effective basis for creating national identity – everyone having a common experience at the same time, building on a common historical experience of national literature and education.

Ironically, mass print media also introduced a new distribution cost, because not only does the information have to reach its audience, but it also has to reach its audience in time. While a book contains information that remains relevant despite any delays in delivery, newspapers and magazines contain information that actually loses its value if it does not arrive on time. Since the information in newspapers goes out of date quickly, the need to transport newspapers physically limits the spatial range of a common readership. Complicated arrangements may exist to ensure newspaper distribution. What is the latest time for production of a newspaper, in terms of train timetables? If the production is late, how will late editions be distributed? Information that is not 'timely' is not subject to such costs, since cheap, even if slow, transportation can be arranged (though all printed materials still have some distribution costs). It is only with recent developments in electronic communication that it has become possible to print the same newspaper in different physical locations, thus permitting some newspapers to enlarge their geographical catchment areas. Before this, transportation costs imposed real geographical boundaries, and local and national newspapers simply did not travel very far. The obvious exception is that of expatriates who read newspapers and magazines containing old news, because the emotional benefit of 'staying in touch' is valuable, even though the news has inevitably been superseded by more recent events.

Modern printing presses, while permitting the production of multiple copies of texts that individuals can afford to purchase, require a significant capital investment in print technology. Unlike speech, in which anyone can produce or receive information, there is an inherent inequality in the mass production of books and magazines. While most people can afford to buy books (especially paperback books) or magazines, relatively few can afford the cost of printing. 'Vanity publishing' is a luxury reserved for the wealthy; most people cannot produce their own information and have little choice about what others choose to make available to the general public. Only a small number of people and institutions are responsible for printing books, magazines, and newspapers. If such books, magazines and newspapers contained only objective and accurate information, such a concentration of information producers would be unimportant. However, there are compelling arguments that the process of producing information introduces distortions and is biased, whether consciously or unconsciously (Curran 1977; Pool 1983; Curran and Seaton 1991). The public have access to only a portion of the information potentially available, and that portion reflects the concerns and perspectives of a small, usually privileged, group.

2.7 Visual and audio information

For many years, information technologies largely focused on text information, even though human beings also use visual or audio information. Text information is not necessarily more important than visual or audio information, but for a long time it was only technologically possible to distribute and store text information. Technologies to manipulate visual and oral information have been slower to develop, even though such information is a crucial component of face-to-face communication, and provides a convincing evocation of places and events that people do not directly experience. Only recently have technologies developed to record and distribute other types of information – especially visual information. Visual and audio information have a greater appearance of 'reality'. Text information is always one step removed – at best, a description of an event, while audio and visual images are seen as duplicating the event, thus creating the illusion of 'being there' at a particular time and place.

For a long time, visual information was like speech – one had to be physically present to see something, just as one had to be physically present to hear something. While writing made it possible to record speech, there was no simple technology to record visual information. A written description of an event was just that – a description of it, not a record of it. Such descriptions could be very evocative, but they were not replicas or records of the visual information. Similarly, one could record a visual image by painting it, but such records were dependent on the ability of the painter to replicate an image on canvas. This was clearly dependent on the skill of the painter, as well as the painter's desire to depict accurately the images, both of which reduced accuracy of the representation. This changed with the invention of photography. Photography, like writing, offered both the promise and threat of removing information (in this case, a visual moment) from its context (Benjamin 1973). One of the first uses of photography was to take photographs of a deceased child, to try to fix, in the parent's memory, what the child looked like. Initially, photography was also seen as a threat to painting, since, unlike a painting, it was an 'objective' record, and so painters worried that people would lose interest in creative, but distorted, views of images. Interestingly, it was the objectivity of the photograph that helped maintain the importance of painting, as the subjectivity of painting added positively to the visual experience (Levinson 1997).[7]

Photography is seen as an accurate representation of real events, without any possibility of human distortion, because it uses a technical process. Photos are, self-evidently, accurate. Recording of speech requires human intervention through transcription (before the invention of tape recorders) and so is a subjective and inaccurate process. In fact, although photography is a technical process, it is not an objective representation of experience because the process can be manipulated. This is something that any photographer knows. A photographer has to frame the picture, deciding what is left in or left out as well as the central focus of the photograph. Even so, one still takes dozens of photos, just to get the right look or appearance. An excellent example is the difference between the photograph of a house that appears in a sales advertisement and seeing the house in person. What is left out of the photo? From what vantage point is it taken? How long did they have to wait to find a day when it was sunny and when cars were not parked in front of the house? The final product is *one* version of reality, yet it appears to be 'the' reality. Visual images are vulnerable to manipulation and alteration while retaining the sense of being 'real'.

As with writing, there was a phase when it was possible only to photograph an event but not to make copies of it. With the invention of the 'negative', multiple copies of photographs, like multiple copies of books, became possible, and such technology soon became widely available. Although there was a time when only the wealthy could afford a darkroom to develop their own photos, an entire infrastructure has developed which permits anyone to pay for their own photographs. For the cost of a roll of film and a cheap camera, negatives can be sent to commercial companies for development, with individuals paying a relatively small amount for the use of the development equipment. However, there remain barriers to the production of multiple copies of such information. Most people have to be content with one or two copies of a print, owing to the cost of making copies from negatives. The only solution is to use expensive production equipment that uses a master version of the photograph to make multiple copies. Thus, it is left for commercial organisations to produce multiple copies for public consumption.

More recently came the enhanced sense of presence of 'being there' that 'moving pictures' enabled. This is a series of still photos is fast succession, fooling the eye into seeing motion. Adding sound, soon afterwards, further bridged the distance between event and viewers. As with the storage of text information, stored visual information could travel distances. Like print, however, there also remained the distribution costs and delays of transporting the physical media from one location to another.

Just as technologies to record and manipulate visual information slowly developed, so did technologies for audio information. Phonograph records and tape recorders were the technological equivalents of photography, replacing the transcription of speech involved in written language. As with print, the problem persisted of first duplicating and then transferring the information. Duplicating became cheaper with the mass production of records, but was restricted to those with access to expensive capital equipment. Transferring also became easier, as one could post an audiotape or a phonograph, but only single copies. Individuals did not have access to mass production facilities. The production of multiple copies of visual or oral information required expensive technology that only commercial organisations can afford.

The development of audio and visual technologies took place only in the last 200 years. However, they are similar to text technologies in that they can be categorised into two modes: the private production by individuals of single copies for other individuals, and the public production by wealthy organisations of multiple copies for wide consumption. There is, in essence, a two-tier system of information distribution. One tier is of private information production in which information circulates amongst a small group and is one to which most people have access. The other tier is that of public information production, in which information circulates amongst the larger public, but which is produced by a relatively restricted group.

2.8 Electronic communications technology

Up to this point, the focus has been on the technologies that record, rather than distribute, information. Information, whether text, audio or visual, that is stored on a physical medium (book, photograph, tape cassette) has to travel from sender to receiver. The cost of distribution imposes significant limitations, on both the cost and speed with which information can be distributed. The invention of electronic communication technologies, such as the telegraph, telephone and, later, radio and television, reduced both the cost of distributing information and the time required to distribute it. Information could be communicated cheaply and almost instantly to any location, as long as the recipient possessed the appropriate technology. These technologies do not suffer from the spatial limitations of previous media, and so altered the social impact of information distribution.

The telegraph was one of the earliest of communication technologies, and, like the Internet, was seen by all consumers as revolutionary when it first developed. This revolution was not in the amount or sophistication of information; Morse code is a labour intensive means of converting text into code (dots and dashes) and then transcribing it back into text. So labour intensive is the process that we all know that a 'telegraphic style' is a style in which extraneous words are dropped (somewhat like contemporary text messaging in mobile phones). The revolution of the telegraph was the speed of information transmission: its ability to transcend constraints of time and space (Standage 1998). Up to this point, the speed of information transmission was limited by the speed with which the physical medium of a book or photograph could be transported; the telegraph transported information at virtually the speed of light or at least at the speed of electric current. The telegraph created a sense of immediacy that had previously been lacking, as time sensitive information ('news') could literally traverse the globe in a day (the only delays being the re-transmission of the message from one telegraph operator to another). The economic value of information was partly dependent on having faster access to it than others have; rapid access to information could make fortunes.

Like the postal service, the telegraph enabled anyone to communicate with anyone else, provided they could pay the appropriate fee. The system depended on an infrastructure to support transmission, and part of the cost of the message was the 'rental' cost of the infrastructure, which not only included the physical infrastructure, but also labour costs such as trained and experienced telegraph operators. The telegraph was reserved for information that was extremely time-sensitive and valuable, but it was available to anyone who could pay for the service. Such messages, while instant, were only single messages, from one sender to one receiver. Like letters, the telegraph is one-to-one communication, rather than mass communication.

The telegraph is electronic communication, but it is not digital communication. The dots and dashes of Morse code may seem to be like the binary code of '0' and '1' that computers use, but they are not. The difference between a dot and a dash is one of magnitude only – one is longer than the other. How long is the key held down to create a long dash as opposed to short dot? It is possible to confuse the two, because the difference is based on length of time, and telegraph operators had to be trained both to create signals that would be understood as dot or dash, as well as to distinguish between the dots and dashes produced by various other operators (all of whom would have slightly different styles). This is analogue, rather than digital communication; the telegraph, like the telephone, radio and television, is based on electric power, but the information that is being transmitted is not coded in the binary form of zero/one.

Just as the telegraph enabled electronic communication of text, the telephone enabled audio communication. It was originally thought that the telephone would be a business tool but it soon became an important factor in social life (Marvin 1988). Virtually anyone can afford the charges for a telephone (which would include the cost of the wider telephone infrastructure).[8] It has had important benefits for people who were geographically isolated, or were isolated by virtue of having to stay at home. It has also become a means for extended families to maintain contact (although international telephone charges tended to inhibit contacts from one country to another). There are limitations of telephone communication that we take for granted. The information is ephemeral, like speech. Once conversation is finished, there is no record of it (except for the later addition of tape recorders to record conversations). In addition, multiple copies are expensive; we do not think of one speaker and many listeners in the context of the telephone.

Radio and television marked the development of mass electronic media. While the telephone and telegraph are one-to-one communication, radio and TV are one-to-many communication. Information producers distribute multiple copies of the same information content to a mass audience. This production process is capital intensive, requiring expensive radio or television production and transmission facilities. The expense of mass production and distribution makes mass media similar to book production, and quite different from the telephone and the telegraph. Even though telephone and telegraph also depend on an expensive infrastructure for transmission, anyone can afford to make or receive a phone call, send or receive a telegram. In contrast, while anyone can afford to buy a book, radio or television set, the production and distribution process for radio and television is capital intensive, which restricts public access to production facilities. There is a fundamental lack of symmetry between producer and consumer in mass-market books, as well as the mass media of television and radio. In addition, the mechanics of electronic communications restrict the number of transmissions (or channels), which further limits public access. There are a relatively few information producers, but a mass of information consumers.

The creation of a mass shared identity was further enhanced by this mass electronic media, especially since the information was ephemeral. People had to listen, and later view, the information at the same time, as there were initially no technologies to record radio or television transmissions. It is for this reason that a contemporary concern, especially after the Second World War, was the proliferation of television and movies made in the United States but distributed throughout the world. There has been an assumption that if people watch television and films produced by another culture, the people themselves participate in that 'foreign' culture, thus leading to cultural domination by United States media. This concern follows from an exaggeration of the power of mass media to change people, but it has become an issue in the global mass media market (and will be further discussed in chapter 9).

Despite the instantaneous character of electronic transmission, radio and television (like newspapers and magazines) are subject to geographical restrictions which limit their audience. Television and FM radio are line of sight transmissions, which restrict reception. Larger areas of reception require additional transmitters and repeaters, involving yet more capital investment. Although medium wave radio is not line of sight, the strength of the broadcast signal also restricts distribution. Radio and television tended to be local rather than global information systems, often national in character. In Ireland, for instance, while some Irish consumers could receive British radio and television before the advent of cable, most Irish consumers received only Irish stations, and Irish broadcasters knew that their audience was largely a national one. This geographical restriction on distribution further enhanced the role of radio and television in fostering national identity.

These restrictions on the distribution of mass media have lessened with the development of cable and satellite transmission. The infrastructure costs for distribution and production remain, but the restriction on the number of transmissions, or channels, has lessened. From the information consumer's perspective, there are now a greater number of radio and television programmes that can be received via cable or satellite, and potentially a greater diversity of programmes available. More significantly, cable and satellite distribution has removed the geographical limitation on audience. There is now a greater possibility of choice, as different people in the same communities have access to radically different broadcasts from different parts of the world. This has encouraged the development of niche markets, as special interest broadcasts that would not have had sufficient audience in a restricted local environment find themselves able to find listeners in a global multi-channel environment.

No longer are people in the same locality necessarily listening to or watching the same programmes, which has reduced the common experience that has been the basis for national identity. No longer can people assume that they can talk with their work colleagues about the television programme they watched the previous night; it may easily turn out that their colleagues were watching completely different programmes. In Ireland in the 1980s, most people would have watched and so could converse about the popular *Late Late Show* shown over the weekend; this was the staple of Monday morning conversations. People can no longer make that assumption, and the pool of conversational topics has been reduced.

On the other hand, local broadcasters can now target international audiences. Irish television can be received in many countries other than Ireland, and television producers can no longer assume a homogeneous, local audience. While cable and satellite may have increased the distribution of outdated US programmes to a global audience, it has also been increased access by the small national and independent broadcasters to larger audiences. National media companies now produce radio or television programmes with an international audience in mind, even while knowing they have lost some of their local audience to competitors from abroad.

2.9 General trends

In this historical review of information technologies, the issues have been technology and social consequences of the costs of information production, distribution, consumption, reproduction and storage. There are information technologies which can provide single copies of information, such as letters, telephones and telegraphs. These technologies are available to most individuals and make it possible for information to be privately circulated within a restricted domain. Then there are technologies that produce multiple copies, making one-to-many communication possible, such as newspapers, books, magazines, radio and television. The cost associated with the production and distribution of multiple copies (essentially, mass media) excludes most people from the production of information. The high cost of equipment also makes it easier to regulate content. If a television studio is destroyed and the broadcasting equipment confiscated, it is difficult to start up again.[9] These two modes of information circulation are not restricted to any single technology. Thus, private individuals can afford only a few copies of their holiday photographs, but commercial publishers can afford the technology to mass-produce them. Individuals can write letters, but making Xerox copies of text is as costly as buying the text commercially, and so is beyond the scope of individuals. These two domains of mass public and restricted private information circulation exist alongside each other.

There are thus two parallel information technologies for the social distribution of information: one-to-one technologies which are available to all and one-to-many technologies which can be accessed by all but can be produced only by a few. Generally, the cost of information mass production has been decreasing, which has led to an increase in the amount of information that can put into the public domain and affordably accessed by consumers. However, the cost of the production and distribution technology has been increasing, so there has been a decrease in the number of organisations that can afford to produce information.

In parallel, there have also been two divergent modes of distribution and storage: information stored on physical media that is slow to distribute, and electronic information that is distributed rapidly but is ephemeral. The former would include books, magazines, letters, photographs and tape cassettes; the benefit of such an information medium was that the information was relatively permanent, but, since it was stored on media that had to be physically transported, there were significant distribution costs. Such information has to be transported from production location to consumption location. If the information is time sensitive (e.g. newspapers), then the transport cost is likely to be higher. Electronic information, including telephone, radio, and television,[10] can be distributed both quickly and inexpensively, but the information cannot be stored. Electronic modes of information also depend on an existing and expensive infrastructure (telephone network, broadcasting network) to transport the information from producer to consumer.

In recent years, new technologies have enabled some convergence between electronic media based information and physical media based information. They have made it possible for electronic information to be recorded and, to a lesser extent, for physically recorded information to be distributed rapidly. The invention of the tape recorder means that telephone calls can be recorded, as well as radio transmissions. Video machines have made it possible for television programmes to be recorded and stored. Fax machines have enabled the electronic transmission of text and photos. These technologies only partially overcome the divergence between the two technologies, however, because duplicates are often of poor quality. A television programme that has been taped is of poorer quality than the original, and a copy of the copy degrades even further. This is also the case for faxed photos or letters, and for audio recordings. In all cases, either the cost of accurate reproduction information remains high, or the quality of the copy degrades rapidly. Thus the asymmetry in the production of mass information remains: high quality reproduction of information requires expensive equipment that private individuals cannot afford.

In addition to the cost of producing and distributing information is the cost of consuming or accessing information, and these consumption costs vary. Books involve only the cost of buying the book itself, and of having learned to read. In contrast, to receive electronic information, one needs a radio or television which has to be purchased and kept in repair, all of which makes the medium slightly more expensive than reading. To state the obvious (because this 'obvious' is now changing), it has also been necessary to purchase separate receivers for different technologies, because information is encoded differently for each media. One cannot watch television on a radio; one cannot carry on a telephone conversation with a television. The consumption technologies are non-interchangeable, leading to a proliferation of distinct artefacts in the home and office.

Overall, since the invention of writing, the cost of producing information has decreased, which has increased the amount of information that can be in circulation. This has also meant a decrease in the 'value' or significance attributed to information before it can be published; it is now affordable to publish information not previously important enough to warrant the production and distribution expense. Distribution and access costs have also decreased, enabling ever more people to have access to information. The 'public' that consumes information has increased to become the entire population. The speed of distribution has increased, and so has the kind of information that can be distributed (e.g. no longer just numbers and text, but audio and visual as well). This has meant that there is continual improvement in the ability of technology to replicate more closely the experience of 'being there', even at a distance, and being able to communicate, at a distance, as though it was face-to-face communication.

Up to this point, none of these information technologies, whether mechanical or electronic, could be described as digital. For many years, there have been significant developments in technologies such as television and radio, including recent advances in satellite and cable transmission, but these are electronic technologies. It is easy to confuse digital information with electronic communication, since both depend on electricity. Information is not necessarily digital just because it is transmitted electronically. Phone calls involve the electronic transmission of information, but the information is encoded as analogue electrical signals (mimicking the pitch and tone of the speaker). Similarly, television was, until recently, analogue information, even though it was broadcast electronically. The transmission of digital information has only become possible in the last few years; for the moment, it requires adaptors to convert the digital signal back to analogue visual and audio information for viewing on televisions or listening on radios.[11] The hallmark of the new information revolution is not simply the new technology of computers, but the digital rather than analogue information that computers process. The digital revolution is changing the characteristics of information production, distribution, consumption, reproduction and storage that have marked the previous history of information technology.[12]

Chapter 3

Information technology and the digital revolution

3.1 What are computers?

Computers are often portrayed as marvellous, intelligent and powerful machines; in fact, they are very stupid, very limited, but very fast, machines that add and subtract zeros and ones. They are restricted to simple yes or no answers to everything. That is to say, they deal with binary data. It may sound impressive to talk of microprocessors which carry out millions of transactions a second, but it is less impressive when you realise that each transaction is just an on/off decision, and it takes millions of these transactions a second to do things that any four-year-old child can do. One of the best-known complaints about computers is that they do what you tell them to do, not what you want them to do. This is because computers can only do what they are instructed to do in a literal fashion; it is programmers who have to devise programs to enable a very stupid, but very fast, adding machine to be the basis of an information revolution.

Human beings process analogue rather than digital information. The difference between analogue and digital information is similar to the difference between old style clocks and newer digital displays. In an analogue clock display, physical movement of the watch hand marks the advance of time. Five minutes is the movement of the minute hand by a specific distance; twice the distance is twice the amount of time that has elapsed. A small change in time is marked by a small, and perhaps difficult to measure precisely, change in distance. Analogue information is encoded in continuously varying values – like varying points along a ruler or on a watch face. In contrast, digital information is based on discrete units; time is represented as either 9:21 or 9:22 but it cannot be between the two. Rulers measure distance on an analogue basis, which human beings then convert to digital numbers. Thus, a distance might be a bit more than 3.12 while not quite 3.13 metres, but has to be written in digital format as either 3.12 metres or 3.13 metres.

Computers are composed of electrical circuits that are either on or off, and they encode all information as digital data in a binary format of ones (circuit on) and zeros (circuit off). It is easy for computers to deal with numeric information; the computer simply converts the numbers we use every day

(0–9) to a series of zeros and ones.[1] Converting letters to digital format is equally simple. There is a standard code for representing letters in decimal form, so the letters ('Jane') are converted to numbers ('74 97 110 101'), which are then converted to binary form ('1001010 1100001 1101110 1100101'). There is no information lost in the transition from alphanumeric (letters and numbers) to binary or back, as the information remains in digital form. This contrasts with the transformation of analogue information into digital form; if the time is 9:21 and a bit, it has to be rounded down to 9:21 or rounded up to 9:22 on a digital watch; the extra bit of time is left out. Similarly, if a distance of 3.12 metres and a bit has to be represented in digital form as centimetres, it has to be rounded down to 3.12 or rounded up to 3.13. Some data are lost in the transformation from analogue to digital format; it is possible to increase accuracy by storing more information (e.g. 3.12567 rather than 3.13), but some loss of accuracy is still inevitable.

The first multi-purpose programmable computers were developed during the Second World War. One important technical innovation since then was the invention of the transistor in the early 1950s, which replaced the vacuum tube. A transistor is still only an on/off switch, but transistors made it possible for a computer to have more on/off switches, using less power. The next innovation was the integrated circuit in the early 1960s, which permitted many transistors to be placed on one piece of silicon. The result was even more on/off switches, for less cost, and with less energy consumed, and greater reliability. Since then, there have been steady increases in the number of transistors that can be placed on the piece of silicon that constitutes a microchip, along with decreases in the cost of producing such microchip integrated circuits. These changes have meant more on/off decisions that can be made per second, at an affordable price, thus leading to increasingly powerful computers. It has led to smaller and smaller computers; but has also made it possible to embed microchips in artefacts that do not even look like computers, such as cars, ovens, microwaves and telephones. Homes that do not possess a microcomputer may still have equipment that functions only because of the microchips that are built into the product.

Human beings exist in a world composed of visual, audio and tactile information, but computers must convert all information into binary form. Numeric and text information can easily be converted into binary form because it is already in digital form, while visual or audio information is analogue information and so is more difficult convert. For instance, the more accurate a digitised picture is to be, the greater the size of the file necessary to store the smaller and smaller dots that represent the photograph. The resulting file is very large, whether to store or transfer; similar problems arise with audio information. It requires great computing power and time to convert the analogue information to digital form, to store and transmit it, as well as to reconvert it back in order to output the audio or visual information. Thus, in the early days, computers were used largely for numeric or text manipulation, leading to descriptions of computers as 'number crunchers'; the storage and manipulation of visual and audio information was beyond the scope of computers.

3.2 The technology revolution

Given that electronic computing has been around for over fifty years, what has happened in recent years that has led to a technological revolution? The information technology revolution is the convergence of developments in two different technologies: computer technology and telecommunications technology. Developments in both, and their convergence into information and communications technologies, have enabled the growth of digital information, with consequent changes in the production, distribution and access to information.

The first thread of technological development has been in computer technology, in which the trend has been for smaller, cheaper, faster and more powerful computers with greater storage capacity. These changes have had real social impacts. Cheaper computers mean more people can have them, and that they are no longer the preserve of large organisations or wealthy individuals. Because computer processing is cheaper, computer applications originally viewed as too trivial for expensive computers can now be carried out. In the early days of computers, word processing was seen as a waste of computer time;[2] now, computers are used in ways not previously possible or economically viable (e.g. microchip-based controls in cars and telephones). More powerful computers have enabled new procedures, such as computer-aided design and computer-integrated manufacturing, and computers can deal with new types of information (such as audio and visual information). A prohibitive amount of time and computing power was formerly required to digitise audio or visual information, now computers can digitise audio and visual information as they receive it, and the digital information can be converted back to sound or pictures equally quickly.

Virtually any information, whether numeric, text, audio or visual can now be converted into digital format, stored or manipulated in that digital format, and reconverted back into analogue form. The increased storage capacity of computers enables the large files created from digitised visual and audio information to be sorted, as well as the storage of vast amounts of text. Computers have now become a consumer item, and are able to process and manipulate information in ways that would have been deemed incredible ten or twenty years ago. Individuals can afford to own computers in their home, and the computers are fast enough to be able to run programmes such as Windows, which are relatively easy to use but which require more computing instructions than character or command-based computer programmes.

In parallel with these developments in computer technology have also been developments in telecommunications. We now take it for granted that computers are linked together and used to exchange information, but this requires a telecommunications infrastructure to link the computers. In the early days of computers, information was transferred from one computer to another via punch cards or magnetic tapes. In recent decades, a infrastructure has developed that permits the transfer of information via electronic data links. The keys to this telecommunications revolution have been two communications protocols and one software innovation. The first is the basis of the Internet: Transmission Control Protocol / Internet Protocol (TCP/IP). This is a protocol (or set of rules) that enables a computer to break information into small packets and send those packets along multiple data links to another computer, whether across the street or across the world, where those packets are reassembled into data. There are many impressive things about this means of linking computers, and one of them is that computers all over the world can be linked, regardless of their operating system. Mainframes, whether made by IBM, Honeywell or Digital, as well as personal computers, whether made by IBM, Compaq, Apple or Dell, can exchange data. This open set of rules, enabling any type of computer to 'talk' to another, has been crucial to the development of a global system of interlinked computers.

A second telecommunications development enabled individuals to access the Internet using their home computers, thus providing an information system that was open to anyone. In the early days of TCP/IP, the only computers that were permanently connected with each other were expensive mainframe computers, usually owned by relatively large organisations. Private individuals could access this network only if they had access to a mainframe in their place of work or if they rented time on a mainframe. In the latter case, they could access the mainframe over the telephone line, using their home computer, but that home computer was not itself connected to the Internet, which severely restricted what individuals could do via the Internet. Then came communications protocols that made it possible for individuals to connect their home computers directly to the Internet via telephone lines. Instead of having to pay for an expensive permanent connection or being dependent on an organisation, individuals could access the Internet directly from home. They could access any information they want, publish any information they wanted, run any programmes they wanted in relative privacy and freedom. Any individual who can afford a relatively inexpensive computer, and can afford the relevant telephone charges, is on equal status with large and wealthy corporations in being able to produce, distribute and access information.

The last element of this telecommunications revolution has been the development of graphical interfaces for computers. Windows has made it easy for people to use personal computers without having to learn and remember complex commands to be typed in at the keyboard. For the Internet, the graphic interface of World Wide Web browsers (Internet Explorer and Netscape) enables people with relatively little computer expertise to utilise sophisticated programmes and protocols in order to access information that resides on computers anywhere in the world. For many people, this last innovation is synonymous with 'the Internet', but it is only one of a number of the developments in computers and telecommunications that have enabled computers to share information.

The scope of this telecommunications revolution has been global; the computing and the telecommunications technologies work in all parts of the globe, and computers can be linked regardless of where in the world they are located. People take this universal scope for granted, but such a global system was not an inevitable consequence of the computer revolution. After all, there are differing technological standards in many things: 110 versus 220 electrical voltage, differing telephone systems and television and video protocols. Despite the differing types of computers, including a large variety of mini as well as mainframe computers plus microcomputers, and despite the diversity of operating systems, the Internet unifies all of them into a single system of peered computers, capable of rapid exchanges of information. Crucially, the information that is being processed and transferred is digital, rather than analogue. This is in contrast to the global electronic – but not digital – systems of television and radio broadcasts and telephone exchanges. The information revolution brought about by the computer revolution has been the content revolution of digital information.

3.3 The digital revolution

The convergence of computing and telecommunications has changed the amount and kind of information that can be digitised, and the way such information can then be produced, distributed, accessed and replicated. The earlier historical review of information and communication technologies drew attention to the capital costs of producing and distributing physically encoded information, as well as to the cost and accuracy of reproducing information. Traditional analogue information is more costly to produce, reproduce, distribute and store than digital information. Secondary copies of analogue information are less accurate, either with new errors being introduced (as a letter is being retyped) or because of an inevitable degradation in quality (as with the second and subsequent copies of a video cassette). Digital information is cheaper to produce, distribute, store and reproduce; as will be seen in later chapters, this changes the social, economic and political 'equation' of information and society.

The telecommunications revolution has changed the nature of mass media. Television, radio and newspapers are 'mass media' because they are produced by a few but consumed by many. The technologies to produce such information are expensive (large printing presses, television studios and broadcast facilities) and this effectively excludes most people from being able to produce and distribute their own television programmes or newspapers. In 1983, a few dozen corporations owned at least 80 per cent of the market for television and radio programming. In 1996, less than ten firms owned about the same share, and many were in joint ventures with each other (Shapiro 1999: 40).

The digitalisation of information has reduced the cost of producing and distributing information for mass consumption. It has increased competition for audiences, since there are more information producers. It has also changed the geographical nature of information flows. The cost of distributing digital information around the world is the same as around the corner, so there are fewer geographical boundaries on information flows. The impact of new electronic mass media is dependent on the emergence of new audiences for this electronic publication. Will people be able to access the 60 instead of six television broadcasts, or 600 instead of 60 radio stations, or the 600 international newspapers instead six national newspapers? Moreover, if they can access them, will they bother? If such access was dependent on consumers having all the separate technologies to receive different transmission media (a cable connection, an Internet connection, a satellite dish, and radio), then few people could afford the cost of such devices and would not bother using them. However, one consequence of the digital revolution has been the overlap, or convergence, of previously distinct information types. Books used to be printed on printing presses, distributed via trucks, and read once purchased, while television programmes were created in television studios, broadcast via transmitter, and received on a television set. A home would have multiple and different information receivers: telephones, television sets, radios, record players and tape cassette players. Now, once information is digitised, it is all zeros and ones, whether that information content is visual, text or audio. It can be distributed using the same technology, and it can be received using the same technology. So the same telecommunications infrastructure can transmit books, radio, television and telephone calls, and the same computer in the home or office can be television set, radio set, telephone, computer and movie projector, all in one.

In theory, this means cheaper information access, because instead of a proliferation of different information devices, only a single device is needed, and this single device (a computer) is getting cheaper all the time. With one computer, and one high-speed connection to the Internet, individuals can access all the digital mass media products: 60 television stations, 600 radio programmes, and 600 digital newspapers. However, things are not that simple. There are often increased consumption costs: the user needs a computer, a telephone and a modem, and these are affordable, but still more costly to buy than a telephone or radio. As the communication content increases in sophistication, there is a constant need for faster computers and telecommunications, as well as a need for constant upgrading of hardware and software. In contrast, the cost of individual items like radio, telephones or books is minor. In addition, multiple items duplicated throughout a house (many phones, radios, TVs) enable information to be received in different locations in the house, or different information by different people (for instance, people watching different programmes on several TV sets in a single house). This is difficult and expensive to duplicate in the digital environment; such sophisticated artefacts are costly to buy and maintain. Indeed, for non-industrial countries, a major barrier to new electronic media remains the prohibitive cost of buying the computer technology and the telecommunications costs of connecting to the Internet in the first place. Access to the multiple digital formats of newspaper, radio, television and telephone remain a luxury for the developed world, unless new technologies can be used to avoid expensive infrastructure investments.[3]

The digitisation of information and the convergence of production, distribution and consumption processes have tended to blur the distinction between the one-to-many of mass media and the one-to-one of private communication. Now a communications infrastructure supports many-to-many communication. Anyone can become both producer and consumer of information, permitting a sharing of information previously unthinkable. New technology also enables more private communication between individuals. Electronic mail, mobile phones, answering machines, digital photos, web cams are just some examples of technologies that are being used to maintain contact between friends and relations, regardless of the physical distance that may separate them. Previous one-to-one communication was restricted in a manner similar to mass media. Letters were inexpensive but slow. Phone conversations were rapid, but expensive. Face-to-face meetings involved expensive travel arrangements. New technologies now allow individuals to stay in contact regardless of location and at an affordable cost.

3.4 Technology and social change

The developments in information technology over recent decades have clearly had profound consequences, and some changes have already been discussed. Most people take for granted that these recent technological changes must lead to economic and social change and the emergence of an information society as a result. In this view, technology is so significant as to be deterministic. The assumption is that technological change leads to inevitable and predictable changes or alterations in other parts of society. This commonsense view is supported by our understanding of the agricultural and industrial revolutions that have preceded this 'information revolution'.

The domestication of plants and animals meant increased human control over the production of food, with a massive increase in the efficiency of food production. Land became an important resource for the first time, and access to land became crucial for survival. Since the amount of land available was limited, various political and legal structures developed to allocate this scarce resource. With sufficient food production came the possibility of sedentary life, since people no longer had to forage in different locations for food. This permitted an investment in fixed resources (things that no longer had to be moved around), and material goods that could be traded, bought, sold or inherited. When supplemented by the technology of the plough and irrigation, a significant increase in surplus food production became possible. When it became possible to produce enough to feed extra people, it then became possible for other people to perform tasks to obtain food in exchange. The exchange might be skills and expertise for food, permitting the development of specialists, such as religious figures, soldiers, scholars and political leaders. Associated with such changes would also be the development of state societies, writing systems, and thus also the accumulation of knowledge from generation to generation. No wonder, then, that one talks about the agricultural revolution. This seems to be a clear case of a technological change in the way food was produced having a profound and predictable impact on the economic and social structure, changing the rest of society beyond recognition.

Similarly, the industrial revolution entailed the development of machines to supplement human labour. This permitted the mass production of goods, leading to capital or money becoming a new basis of wealth, in addition to land and labour. The change in production permeated society, changing social relations, and where and how people lived. Cities grew, as people lived near industrial centres, and new services developed to cater for urban living. Thus we see a set of technological changes with economic and social consequences. There is some ambiguity in the causal relation, since it has been argued (Weber 1958) that there were some ideological changes that had first to take place, before the emergence of industrial societies was possible. Usually, however, it is assumed that technological changes lead to economic and social changes.

Small wonder, considering the technological revolution described in this chapter, that many people expect far-reaching economic and social changes. Certainly the level of technological change has been dramatic and visible. There have been breakthroughs in information processing, storage and transmission, which have led to the application of IT in virtually all sectors of society. The reductions in cost and increases in power of computing have meant the proliferation of embedded microchips everywhere and in everything. One can hardly use any artefact that does not have a small microchip in it. However, while the dramatic sweep of technological change is clear, the economic and social consequences are less so. If Irish Internet access rates for adults (from age 15 up) were 33 per cent in 1999, up from five per cent from three years earlier (Information Society Ireland 1999), and up from 16 per cent one year earlier, does that mean Ireland is now twice the information society it was one year earlier? Since access to a PC has doubled to 41 per cent from 1999 to 2000 in Ireland, is Ireland now twice the information society it was a year ago? Does it matter if there are disparities (gender, education, region) in usage and ownership that are concealed by these statistics?

Measures of computer ownership or usage are meaningless on their own. Having computers does not necessarily mean anything; for example, having computers in classrooms means little unless they are used for something. Technology measures are attractive because they are quantifiable, but it is not how many are online or how often that matters; it is what people are doing when they are online that is important. How is that usage changing their lives in other ways? Is there less socialising with friends? Do they stop going to shopping malls? Are they better informed? As more parents find their children working outside Ireland, will they learn to use email to keep in touch? Will they learn to use the World Wide Web to see instant digital photos of their grandchildren? These are more important social changes than measures of the number of computers in a community or country, and the reason why technological measures are meaningless, if taken in isolation from their social context. Having a computer does not necessarily or inevitably change a person's life; the discovery that more people have computers cannot be used as a predictor or indicator of social or economic change.

3.4.1 *Technological determinism*

Technology has an impact on human behaviour, but it does not determine human behaviour. For many of us, technology seems self-evident and its impact is non-problematic; as new technologies are developed, they have a clear impact on economic and social life. So government policy may encourage investment in high-speed communications links on the assumption that economic benefits follow inevitably and predictably. In fact, the relation between technology and society is complex, and is now studied in the context of social studies of science and technology (Fuller 1997; Webster 1991; Yearley 1988; Woolgar 1988). There are two extreme views of the relation between technology and society. One is a deterministic view, suggesting that technology has an inevitable and predictable influence over society. The other view sees technology as neutral, with society determining how technologies will be devised and used.

Within the deterministic camp, there are both optimistic and pessimistic views regarding the impact of new technology (Kling 1994). The optimistic view has a distinct utopian flavour. Any social problems can be fixed with technology, and technology will overcome social evils, essentially going towards a world in which technology has provided food to alleviate hunger, cured both physical and mental illness, and improved the standard of living for all. In this future, information is easily available and this availability is used to enhance democratic participation. Technology seems part of an inevitable progress towards a better future. This view was especially strong in the mid-twentieth century but continues to this day, as reports of new discoveries or inventions herald solutions to a range of problems. This utopian view of technology has become somewhat tarnished in recent years. We now know that the consequences of technology are not always beneficial, nor is progress inevitable. There are problems that technology and science cannot solve, and that solutions have unintended and harmful consequences. In addition, technology and science can actually create problems: nuclear energy creates pollution, new drugs encourage antibiotic resistant super-bugs, industrial production leads to acid rain and global warming.

Opposing the utopian view is a more pessimistic view, in which technology is used to control and monitor individuals. Information is denied, access to information is controlled, and individuals are under surveillance and lose their privacy. Technologies are used to maintain existing inequalities and enhance them. The superior position of states and multi-nationals vis-à-vis the individual is enhanced, rather than diminished, as they use their superior access to technology to maintain their position. The image of George Orwell's 1984 was a profoundly strong one, in which governments decided what was true, rewriting history to suit political purposes, and surveillance technologies enabled governments to reduce individual privacy so that individuals could be monitored and controlled. This pessimistic view is also extreme. While personal privacy has diminished in recent decades, recent history provides instances where technologies have been used to enhance the political power of the marginalised. Recent examples would include protests against the World Trade Organisation, organised via the Internet. Information has become a political issue, and there has been a growth in data protection legislation and freedom of information legislation to redress the balance of power between individuals and organisations.

These optimistic and pessimistic views share a belief in the primacy of technology over individuals and society. Both views share an assumption that there is little that individuals can do to alter the irresistible force of technology. Individual free will and choice have little place in these views of technology: individuals can accommodate or be overcome.

3.4.2 *Technology as neutral*

Not everyone agrees that technology is outside human control or that it determines economic and social change. An alternative view is one that sees society as more central, and technology as a tool to achieve socially desirable ends. In some cases, the ends may be consciously determined, but in many cases the ends are neither articulated nor actively decided. In this view, any technology (including information technology) does not itself have an impact; it is 'neutral' and its impact is the result of human action. Sometimes this is primarily used to justify non-intervention. Scientists, for instance, would say that they simply discover or invent nuclear bombs, *in vitro* fertilisation, biological weapons, or contraception, and then it is up to society at large to decide how they are to be used. The clearest example of this is the slogan of the National Rifle Association in the United States: 'guns don't kill, people do'. This slogan is an explicit attempt to resist gun control regulation by arguing that technology is outside society (Shapiro 1999: 13). Of course this slogan can also be turned on its head and interpreted to mean that it is society's duty to regulate guns by banning them. Or that it is society's duty to regulate biological weapons by banning research into them.

Technology can facilitate or impede social behaviour, even when not a determinant of behaviour, and so technology can be designed and implemented in ways that achieve social or political goals. Sometimes these goals are political in a global sense. Telecommunications research in the 1950s aimed to help scientists use expensive computers in the most efficient way; that research was funded by the United States Department of Defense and the scientists benefiting from the research were devising better methods of warfare. Artificial intelligence research was to improve military decision-making and to replace soldiers with computers. Hardware developments were funded so that small computers could be sent up into space. Much of the technology that has underpinned the 'information revolution' was funded by governments who wanted to use the technology for military aims (Roszak 1994; Robins and Webster 1999).

However, technological innovation also results from political agendas that are 'subversive'. One aim of those who developed early personal computers was to liberate individuals from control of large corporations like IBM. Mainframes were seen as imposing a centralised control, reducing individual freedom, and microcomputers were to provide low cost alternatives that would free people from this control.[4] Early developments in networked communicating, such as FidoNet and UseNet, were the results of individual computer programmers who wanted to be able to share information and programmes. The best-known examples of such innovation are the World Wide Web, which was designed to facilitate the sharing of information, and the PC version of Unix (Linux) that is a freely available counterbalance to the Microsoft Windows operating system.

Since the demise of the Cold War, research funding is more likely to have an economic agenda than a military or political agenda. Current research in encryption technology will enable secure economic exchanges over the Internet, and there is a significant amount of technological research in the application of copyright and intellectual property laws to digital information. There is also technological innovation designed to achieve social ends. For instance, technology that will monitor individuals who live alone and may need medical intervention or communications technologies that will make it cheaper for individuals to access the Internet. Whether the agenda is military, political, economic or social, there are numerous examples of technology developed to achieve specific ends. Thus, to argue that technology is outside human control is simply to cede that control to other interest groups in society.

Designing technology for specific social goals is often problematic. Even when technology is designed for specific purposes, the impact of technology on society is never clear-cut or predictable. A technological innovation or discovery is accepted for reasons that often have little to do with the technology itself. As already noted, the success of VHS over BetaMax was a result of marketing – not technical – issues. Apple never permitted Macintosh clones, whereas IBM tolerated PC clones and could not restrict the use of MS-DOS software. The price of Macintosh hardware remained prohibitive, whereas IBM PCs were quickly copied, became affordable, and as an increasing amount of inexpensive software was written for inexpensive IBM clones, IBM PCs became dominant. The communication protocols of the Internet (Transmission Control Protocol and Internet Protocol) were published by Robert Kahn and Vincent Cerf in 1974, yet the Internet only became pervasive in the late 1990s (for a detailed history of the Internet, see Hafner and Lyon 1996). The history of technology is replete with such examples. In the early days of refrigerators, there were silent gas fridges and nosier, compression driven, electric fridges. The noisy electric fridge is now the standard fridge. Is this because it was better and cheaper, despite being noisy? No, it was better marketed and distributed (Cowan 1983: 127–45); economic, not technological, factors determined its success.

What is the cause and effect between technology and behaviour? If a husband leaves his wife after email contact with another woman, did technology cause the separation or did the technology merely facilitate a pre-existing social process? On the 'cause' side of the argument, electronic mail is private, so the husband was not observed. Email exchanges can be more intense than face-to-face communication, since there are few distractions when individuals are involved in electronic communication. Relations can be started without risk, since most such exchanges are anonymous (at least at the beginning). Thus a relationship developed when it would not otherwise have had a chance, and exaggerated any dissatisfactions that already existed in a relationship. If it were not for the technology, the couple would still be together. On the 'facilitate' side of the argument, the Internet is only a technology; people still have free choice. The husband must already have been unhappy in the relationship, or else he would not have taken advantage of the opportunities which chat rooms and email enabled. Even if there had not been the Internet, he might have become involved with someone else anyway. Or, even if there had been no other relationship, he would have been quietly unhappy. The technology only allowed a pre-existing problem to be expressed, and one should not shoot the messenger. Which is the true explanation? In fact, there is no clear unidirectional relationship between technology and individuals' lives.

3.5 Technology and the information society

The significance of the revolution in information technology is undeniable: a digital information revolution has taken place. The economic and social consequences of this revolution may also be significant, but they are less clear. Although some examples of the impact of technology on society have already been given, it is difficult to predict the social impact of technologies. Their impact is not uniform, since the same technologies have different impacts on different societies. The relation between technology and society is unclear; there are positive or negative impacts that the same technologies can have on society. This depends on how societies determine the function and role of technologies, since social forces can alter the design and adoption of information technologies. For that matter, the same technologies can be subjugated to different economic and social goals in order to achieve quite different outcomes. We can take the information technology as a necessary but not sufficient condition for social and cultural change; the interplay between technology, economy, and society is far too complex for any unidirectional causal models of technology altering society. One must also consider the other factors that are relevant in this digital information revolution.

Chapter 4

An information economy

Economic change has been the most visible consequence of technological change and innovation. It is often assumed that technological change leads to economic changes, and it is those economic changes that dominate discussions and descriptions of the information society. Changes in the way we work, the kinds of jobs we do, the cost of things we buy, and other economic changes have a profound impact on people. It is economic change that seems the most direct and inevitable consequence of technological change. However, since the consequences of technology are neither predictable nor inevitable, it is necessary to look more closely at the economic changes that loom so large in the minds of most people. These changes are driven by technological changes but by other forces as well. Will changes be profound enough to see the emergence of new economic forms, and will these new economic forms be linked to wider political, social, and cultural forms?

4.1 A post industrial society

Because of recent developments in technology, many argue we no longer live in an industrial society. Previously, the economic basis of society was manufacturing, and especially mass or assembly line production. Technology has so changed the production, distribution, and consumption of economic goods and services that 'industrial' is no longer an appropriate description of economic life. The view, most closely associated with Daniel Bell (1973), is that information has become the central economic resource, replacing the land, labour and capital that underpinned manufacturing and industrial production. It is often assumed, though not always clearly articulated, that this economic change is so significant as to change the entire fabric of society, altering social, political and cultural relations as well. This new economic basis of society has variously been called a 'post-industrial society', an advanced industrial society, or an 'information society'.

This view is based on a few central propositions about the structure of societies. Firstly, the proposition that economic activity can be divided into three sectors: (1) a primary sector, including agriculture, fishing, mining and other extractive sectors; (2) a secondary sector, composed of construction and other manufacturing industries; and finally (3) a tertiary 'service' sector

(Clark 1940). The service sector provides non-material outputs, the sort of outputs which lawyers, doctors, civil servants, hairdressers and sales personnel produce. Secondly, one can characterise an economy by the relative importance of the different sectors. For instance, in an industrial society, agriculture and fishing as well as service activity also exists, but industrial production would be the most important sector and would also permeate the other sectors. In an industrial society, agricultural productivity would be based on manufacturing techniques. Thirdly, productive economic activity is moving from one sector to another over time. The industrial revolution was a move from agricultural activity to industrial activity, and there is now a move towards service activity, with industrial production diminishing in importance. The dominance of this new service industry, and especially the growth of information related activity within the service industry, is leading to the emergence of a post-industrial or information economy. The final proposition (and perhaps the most problematic) is that an entire society can then be characterised by its central economic activity. Thus an industrial economy leads to an industrial society.

It is a relatively easy process to classify economic activities in the three sectors of primary (agriculture, fishing, mining, etc.), secondary (industrial production), and tertiary (service industries). There is also good historical evidence that the industrial economy involved a move from primary to secondary economic activity. Agricultural employment has been declining since the turn of the century while manufacturing employment has been rising; in the United States, agricultural employment dropped from 87 per cent in 1800 to 35 per cent in 1900, and down to 15 per cent by 1940, while industrial employment rose from one per cent in 1800 to 26 per cent in 1900, and 37 per cent in 1940 (Beniger 1986: 23–4; see also Machlup 1962; Porat, Rubin et al. 1977; Stehr 1994). Furthermore, agricultural activity has become 'industrial' in character, as evidenced by mechanised farming, fertiliser intensive farming, and large land holdings, with output sold on the world market. The value-added component of agriculture has been the introduction of industrial processes in agriculture. Therefore, the industrial economy can be defined by the importance of manufacturing, but also by its impact on other economic sectors as well.

The post-industrial society, it is argued, is marked by an increase in service activity (transportation and public utilities, wholesale and retail distribution, finance, real estate, insurance). Just as manufacturing characterised the new industrial society, 'service' employment identifies the new post-industrial economy. In pre-industrial societies, manual labour was the pervasive type of work; in industrial societies, people worked at machines; white-collar service employment defines the post-industrial society. Information-related work ('paper work') becomes increasingly important, with the emergence of an 'information sector' that permeates the other economic sectors such as agriculture and manufacturing.

Daniel Bell is most closely associated with this argument, publishing *The Coming of the Post-Industrial Society* in 1973. This publication corresponded with major economic changes in the early 1970s, as microelectronic processors began to permeate and transform offices, industrial production, schools and homes. There was the prospect of computers doing tasks that previously were either done by human labour or, more importantly, had not been previously possible at all. However, the crucial element of Bell's work was to argue that this economic transformation was also a social transformation. New technologies were altering the very basis of society. He especially emphasised the impact of science and technology on society:

> In Western society we are in the midst of a vast historical change in which old social relations (which were property bound), existing power structures (centered on narrow elites), and bourgeois culture (based on notions of restraint and delayed gratification) are being rapidly eroded. The sources of the upheaval are scientific and technological (Bell 1973: 37).

Bell explicitly linked technological change, economic change and social change: technology caused economic change and economic change leads to social change. With automation and mechanisation would come job losses in agricultural and manufacturing sectors. However, growth in the service sector would mean more jobs, to soak up surplus unemployment. This change would also mean more wealth, since greater productivity would mean less work was needed and greater profit was obtained. People would use this profit to purchase things and services that they could not previously afford. As people's material needs were satisfied through increased manufacturing output, their desire for intangibles such as health and education increased. The desire for social benefits reinforced economic changes, thus creating a post-industrial society.

In Bell's vision, an important element of this post-industrial society is a change in the basis of power and wealth. After the agricultural revolution, land and labour were the important economic resources, and political power derived from control over such resources. With the industrial revolution, capital (owning a factory, for instance) became the basis of wealth and power. With the growth of white-collar and knowledge-intensive work, scientific workers and knowledge workers would hold increasing power. Knowledge, rather than land, labour or capital would become the 'axial principle' of society. Knowledge would become the 'added value' of an economic activity that enhanced its profitability. One could make a certain amount of profit from land or from capital, but eventually there comes a point of diminishing returns. More land, labour or capital would not dramatically improve productivity nor would it provide sufficient return on investment. However, knowledge could make land, labour or capital more profitable. For instance, one could farm more productively with fertilisers and seeds particularly appropriate for soil conditions and climate. Scientific research has been used to produce better fertilisers and new seeds, and farmers have the knowledge to know which products are best for their particular soil and climate. Weather forecasting enables crops to be harvested at the best time; improved storage technologies mean that the harvested food can be shipped greater distances and sold in new markets. Thus crops can be planted to provide the greatest profit with least cost. There is increased dependence on documentation, as agriculture becomes part of the wider bureaucratic process with farmers having to fill in forms to obtain government payments. They now have to demonstrate a disease-free herd by having microchips embedded in cows in order to trace their movements. Information technologies also lead to more efficient factory production, with less spent on labour, less capital tied up in inventory through a computer-based inventory system, and a faster and cheaper turn-around time for producing new goods. Thus, the post-industrial economy is marked by information becoming the source of wealth and power.

Bell argued that these economic changes would have a fundamental impact on society, leading to the emergence of new social groups defined by non-economic criteria (e.g. 'post-materialists', 'new working class', 'technocrats'), the distinctive importance of information (leading to more creative and satisfying work), and a decline of traditional class conflicts. These changes would lead to the emergence of a consumer society, in which people no longer need to satisfy economic needs, and would thus look to satisfy non-material needs. The new post-industrial society would be a far nicer, as well as a far wealthier, society in which to live. This is a view of economy and society in which progress is inevitable and change is beneficial.

4.2 An advanced industrial society?

Bell's predicted post-industrial society has been accurate, at least about the broad sweep of technical and economic change. There has been in increase in service industries (not necessarily at the cost of manufacturing employment, but probably because of decreasing agricultural employment). There have also been major changes in manufacturing, in response to changing consumer demand and a saturated market. It is no longer possible to produce the same goods for the same consumers; one must have short flexible production runs, with new products emerging quickly, often targeted at niche or narrowly defined markets. Profit margins have decreased, and competition has increased, with information providing the value-added profit. We live in a consumer-led market society and people spend large amounts of money on consumer items such as foreign travel, clothes or technical gadgets.

Another aspect of Bell's formulation about which there is little doubt is the primacy of scientific knowledge in economic life. Information has an economic significance that attracts enormous investment in research and development. Scientific discoveries are transforming daily life, and such knowledge has an economic value. Biomedical and DNA research, for instance is an economic, rather than just scientific, activity. Information is a commodity to be bought, sold, and owned, and is protected by legislation, in the same way as property is protected, with intellectual property and copyright legislation updated to reflect these changes. Information is a valuable economic resource, often of greater value than land, labour or capital.

However, not all aspects of Bell's vision of economic change have been clearly evidenced (Robins and Webster 1987; Stehr 1994). For instance, technological innovation has reduced labour requirements in industrial and agricultural sectors. But there have also been similar innovations in the service sectors. Much of the recent transformation in work is towards customer self-service (automated teller machines, ordering goods and services such as computers, books and airline tickets over the Internet) which is changing the nature of the service sector (as travel agents, book shop personnel and bank tellers become redundant). Growth expectations may not be fulfilled; despite technological investment, there have been slow rates of growth since the late 1970s (Preston 2001).

More importantly, doubt has crept in about the social consequences that Bell expected from a 'post-industrial society'. As the nature of work changed, those who lost their jobs in old industries were supposed to be absorbed in new industries. Instead, there is now sectional unemployment: agricultural and manufacturing workers who have lost their jobs are not able to retrain sufficiently to move into the service sector and so remain unemployed. The new service jobs have, instead, been taken by the recently educated and young. The result is a serious risk of a permanently unemployed underclass.

Bell also expected positive improvements in the conditions of employment, the nature of work, and the profit derived from work. Work was to become more interesting as well as more financially rewarding. In fact, just as the assembly line manufacturing process reduced the skills needed to produce goods, so new technologies are reducing the knowledge and skills necessary for white-collar work. There is now significant de-skilling and routinisation as new technologies reduce the contribution of individuals to the work process. Instead of engaging in interesting and demanding intellectual work, many 'information workers' are little more than assembly-line workers; they act as extensions of computer programmes, carrying out actions dictated by video display units. Cashiers at a supermarket follow the instructions on their cash register; service personnel responding to customer queries follow the instructions on their computer screens. Even the pivotal position of science and technology is more complex than originally believed; the new 'information worker' can sometimes be as poorly paid as his or her 'industrial worker' predecessor. Most information workers are still wage earners in a market economy and their wages fluctuate, depending on supply and demand.[1]

These doubts about a post-industrial society have been articulated by theorists who do not see any radical changes in society, despite the economic changes that have been taking place. Somewhat along the 'rich get richer and poor get poorer' argument, these theorists argue that conflict and inequality are fundamental to society, and new information technologies have not changed such fundamentals. Alain Touraine, for instance, has argued that a new working class is being formed out of scientific and technical as well as industrial workers (Touraine 1974). Information and communications technologies enable employers to dispense with the skills and expertise of white-collar workers, making them the functional equivalent of assembly-line workers. With bar code readers, there may not even be a price on an item being purchased; cash registers now maintain information about prices, and the checkout people do not even know prices of goods or when they change. Cash registers can instruct workers what to do next (e.g. 'check for credit card signature'). Workers may not even need to read English: pictographs of hamburgers or fruit may be used if barcode readers are not being used. The employment conditions in call centres, with individuals in cubicles and answering the phone for hours at a time, do not correspond to Bell's positive vision of work in a post-industrial society.

This is a return to the previous argument about technological determinism. Technology is not having an inevitable (in this case, positive) impact on society; interest groups are appropriating technology to maintain existing power conditions. Herbert Schiller would argue that the same old questions have to be asked, whether it is new technology or old technology: for whose benefit and under whose control will new technologies be implemented (1973: 175)? To what end, and with what consequences, is information technology being introduced and the information sector expanding? The commodity may be information, rather than labour or capital, but this does not necessarily alter the power relations between worker and employer. Even Bell's primacy of science and technology is a distortion, in Schiller's view. There may be investment in development, but it is targeted at developing market products, to be sold for profit. Research funding is aimed to provide market advantage, as opposed to being for the sake of 'pure' research. Thus, Schiller would argue:

> What is called the 'information society' is, in fact, the production, processing, and transmission of a very large amount of data about all sorts of matters, individual and national, social and commercial, economic and military. Most of the data are produced to meet very specific needs of super-corporations, national government bureaucracies, and the military establishments of the advanced industrial state (1981: 25).

Knowledge and technology are advancing the goals of specific economic groups, which is a rather different view from the one articulated by Bell. In this alternative view, economic processes remain much the same, and, although a new type of economy has emerged, the social and political forces of industrial society remain in evidence. Thus, this is not a new information society, but simply an advanced industrial society; the commodities have changed but the divisions between rich and poor remain.[2]

4.3 A manufacturing transformation

Whether new technologies have led to a post-industrial society, an information society or an advanced industrial society, there is no doubt that new technologies have created a new economic sector, transformed existing sectors, and altered the nature of employment. As already noted (see pp. 51–2), the information revolution has transformed agriculture. The changes in the manufacturing industry have been even more marked. Assembly line production previously meant that economies of scale were achieved by producing large numbers of the same item, year after year. Manufacturers decided what the consumer wanted, and then produced large quantities of it. This assumed that there was an undifferentiated consumer market in which people bought the cheapest item available: there was little choice in supply, and little variation in demand.

Since the Second World War, consumers in Europe, North America, Australia and some Asian countries have had increasing amounts of disposable income, and have used that extra income to purchase consumer goods. As consumer spending increased, the existing market became saturated, so manufacturers needed to sell new goods to those who already owned goods. Producers became inventive, and one strategy has been the emergence of specialist or niche markets (the sports market, the under-21s market, the male-oriented gadget market). Alternatively, manufacturers bring out products that have new features making the previous product obsolete. In addition, goods are produced on a 'disposable' basis: they are designed to wear out rather than be repaired, thus requiring consumers to replace them with new goods. Finally, new versions of products are constantly introduced, each version having slight innovations that would encourage people to get rid of their existing radio to buy a new radio. Or a change in style and fashion would encourage people to buy a new jacket or shirt, replacing existing clothing that had become unfashionable.

Complex and differentiated markets emerged, and manufacturers needed to produce new and varying products with great rapidity. With increasing competition, they also needed to reduce overheads and increase productivity.

These changes were made possible by the use of information technology to introduce significant changes in design and production. Technology improved productivity, which then enabled reductions in labour expenditure. It also permitted more economical allocation of existing labour by means of new control mechanisms. It even altered the actual production process. New production systems, with embedded technology, make changes in manufacturing procedures possible without having to make new investments in expensive machinery. This permits short production runs, with new products available quickly and without enormous capital investment in new equipment. New technology has enabled faster designs of new products, so that organisations can produce small numbers of goods for niche markets, and quickly change to producing different goods for other markets. This transformation has not been without problems. In addition to labour problems caused by reductions in the workforce, the new machinery requires a large capital investment. With increasing competition and decreasing profit margins, the profit to repay the capital investment is sometimes slow in coming. However, regardless of any problems for particular organisations, the overall change marks a sharp break with previous production systems.

The strongest image of manufacturing during the twentieth century has been assembly-line production: large plants turning out a large number of the same product, with individuals doing the same specific task day after day. This has generally been labelled as 'Fordist' production, taken from the Ford motor company assembly plants. Current changes in production have been described as post-Fordist production. Post-Fordism describes the emergence of flexible specialisation and globalisation in the 1970s, where production processes are modular and are distributed throughout the world. Goods are sourced from one location, partly assembled in another location, further assembled in another location, and then sold on the global market from yet other locations. This is possible only because of information systems, linked by global telecommunications, which permit the co-ordination of production, the planning of distribution, and global money exchanges.

Another hallmark of post-Fordism is flexibility. This involves the flexibility of employees, who no longer have rigid and unchanging job descriptions. There is also flexibility in employment, as the tendency is now for people to move from one job to another, working as a contract labourer rather than as a 'company' person, and increasingly to be employed in flexi-time, telework and job-sharing arrangements. With new technologies, short production runs are possible, making it feasible to produce a greater variety of products, aimed at smaller niche markets of consumption, without incurring major capital investment costs. This flexibility has led to a shift away from mass-production, with a consequent increase in the diversity of goods, aimed at smaller and smaller groups.

This increased choice in consumer goods is what people are most aware of when experiencing 'post-Fordist' production. Mass electronic goods, like TVs or mobile phones, are disposable in the sense that people get rid of them not because they do not work but because they do not have some of the functions now 'necessary'. A demand is created by providing new functions or benefits for which new products then become a necessary purchase. Often, if a product fails, it is simply cheaper to buy a new product than to fix the existing one because a production process designed to produce goods cheaply also produces goods that are difficult to repair. This new consumer world is only possible because of new technologies in the design and production process.

While new technologies have seen changes in the production of existing goods, and the production of new consumer goods (mobile phones, computers), not all products lend themselves to machine production. The value of some products lies precisely in their creation by human beings and the very fact that no two products are exactly the same. Waterford crystal must be blown and etched by individuals, and the fact that each one is unique by virtue of individual labour gives them added value. Hand drawings are more valuable than mass-reproduced drawings, hand-carved and hand-painted goods are more expensive than machined goods, all precisely because of the human labour. Then there are new 'cultural' products, such as music, films, plays, and indeed new industries such as the tourist industry, which are packaging cultural heritage and selling it as a commodity. This is directed not only at people who visit other locations as tourists, but also people who buy goods such as food and clothing which are part of that cultural 'product'. Aran jumpers as well as farmhouse cheeses are all part of this new production process. These new 'cultural' commodities were predicted by Bell when describing the post-industrial society: there are machine-produced products whose value lies in utility and there are hand-produced products whose value lies in their symbolic value.

4.4 Information work and organisational change

These changes can also be detailed at the level of the worker and organisation, as information has transformed the extractive, manufacturing and service industries into activities that are information intensive and computer dependent. These changes alter the structure of organisations, relations between worker and employee, and the nature of work. All organisations routinely generate and then process far more information than ever before, leading to the accumulation of enormous storehouses of data on individuals and products, and a move from paper to electronic documents. Far more of the data are analysed, in far greater detail than ever before, leading to the creation of complex systems of administration and decision making. New information and communication systems now make dispersed work groups possible, where members may work in different branches anywhere in the world. More significantly, it also enables more centralised control of operations in branch offices by central headquarters, with less autonomy for managers in local offices. It has also altered the relationship between suppliers, customers and businesses: suppliers often receive orders electronically from an organisation, just as customers often input their own purchase orders via the Internet.

All occupations are affected by these changes. In some cases, existing jobs may require only additional skills. Even jobs that cannot be easily automated (e.g. farm labour or fishing) still require new skills or new tools. For instance, fishermen now use satellites to check their location. In other cases, the job retains the same name, but the job description changes as new tasks are expected of workers. For instance, the occupation of 'secretary' continues to exist, but its job description has changed radically, as secretaries now carry out tasks that would have been someone else's responsibility a generation earlier. In addition, of course, new jobs are created, and old jobs simply disappear. As tasks are automated, as productivity benefits permit workforce reductions, or as cheaper labour can be obtained somewhere else in the global economy, the threat of unemployment is inevitable. Will people retrain in order to qualify for jobs that did not previously exist? Or will they remain out of work (perhaps too old to retrain, or just unable to obtain the relevant expertise), with the new jobs taken by young, recently trained, graduates of a third-level institute of education?[3]

The changing requirements of work have led to a changed attitude towards experience and training. Years of experience formerly increased one's value to the organisation, and seniority carried more responsibility and prestige. Now, previous experience is no longer always relevant or valuable for the new tasks required. Often newly acquired expertise is required, and the skills of existing employees may be of decreasing use to the organisation. Unless staff retrain, prestige may now decrease with seniority. Instead of 'first in, first out', it may be the senior staff who find their jobs under threat. For those who remain in employment, a job is no longer a matter of mastering an unchanging (or only slowly changing) set of practices and skills. Employees constantly have to acquire new skills and expertise.

Employees can make themselves more valuable by obtaining additional expertise and experience; the unresolved issue is whether one acquires this expertise with the assistance of one's employer, or on one's own. Companies must be flexible and adaptable, changing as the market changes, which requires new skills and new jobs. Rather than employ people with those skills, and constantly retrain them as requirements change, companies sometimes employ people on short-term contract. Companies buy in the latest skills and expertise, use them for a short-term project, and then dispense with them, keeping only a core of permanent employees. It is often assumed that the new information economy means the end of 'jobs for life' as long-term permanent employment is replaced by skills based short-term contract employment. There is evidence that such an orthodoxy may be an overstatement which reflects assumptions rather than statistics; long-term employment in the United Kingdom is increasing rather than decreasing, although it does seem linked to skill levels (Doogan 2001).

Regardless of changes in the nature of employment, skill levels remain crucial for employment. Where do employees gain this expertise? In some countries, such as Ireland, companies have been able to influence government educational policy so that the educational system provides the skills which companies require. This is obviously advantageous for companies, who acquire expertise without having to invest in training. However, it encourages companies to 'update its knowledge base by simply recruiting new graduates, rather than by re-training existing staff', which has been described as a 'slash and burn' relationship to the educational system because it uses the products of the educational system without developing human resources (Wickham 1998a). This leaves individuals responsible for developing their own expertise. Small wonder that there has been such a rapid expansion of 'lifelong learning' programmes, so that people can acquire new expertise that either keeps them employed or enables them to compete on the open market for new employment.

For many employees, new technologies have also reduced their autonomy, by enabling knowledge to be embedded in information systems, which has led to greater automation and less individual discretion. The criteria for decisions are determined by the computer system, as are the options available in any given situation. Work in high tech offices can be as tedious, high-paced and stressful as old assembly-line jobs. Furthermore, as people need to know less, they become cheap and replaceable, leading to reductions in wages and worsening working conditions. Staff turnover increases, and few employees have overall knowledge of the organisation or any commitment to the long-term success of the organisation. In short, white-collar work begins to resemble the mass production assembly lines that are emblematic of the industrial revolution (Aronowitz and DiFazio 1994; Braverman 1974).

Of course, new technologies can also enable new conditions of employment that benefit employees. For instance, new technologies permit individuals to work from home. This is not new, as managers and white-collar workers have often taken work home. However, telework means that individuals who work at home have access to all of the same information and individuals that they would have if they were in their office. They can share files, access databases, communicate with colleagues, and engage in any other activities they would normally carry out in the their place of work. This benefits individuals, who may want to reduce commuting time, combine work with household responsibilities, or live in a rural area removed from the city. It also benefits companies, since telework reduces costs such as heating and office space rental. It may also allow companies to hire specialists who would not want to relocate to the actual employment site or for whom the company could not afford to pay relocation expenses.

Managers are often unenthusiastic about telework because it is difficult for them to supervise work and ensure productivity. They prefer to check when and how long an individual is working, and such monitoring is impossible in telework. One solution is to monitor work levels by monitoring output, so that workers achieve an agreed level of productivity. This is a time-honoured mechanism by which employers encourage productivity by rewarding hard work and replacing external managerial coercion with internal self-motivation: 'piecework'. However, piecework is often associated with exploitation of sweatshops in poor urban areas, and, if not regulated, can lead to electronic exploitation of those who work in isolation in homes. While some teleworkers are highly paid specialists, many are employed because of their willingness to do a job for less pay. They are not in a strong bargaining position, and may find themselves isolated and poorly paid. Often, these are not only rural (and so can manage on less pay than their urban counterpart), but also female (Webster 1996a).

Whether the employment location is 'in-house' or external, technology increases the amount of surveillance in the workplace and monitoring of productivity. Since so much work is now carried out via computers or machines with embedded computer chips, it is relatively easy to measure the output of workers. How many customers pass through each checkout counter, or how many computers are sold by each telesales employee? Such information may be used to determine wages and bonuses, as well as to keep track of overall business trends. Technology can also monitor activities in the workplace: cameras can videotape work, phone calls can be monitored, and email and web users can be identified and tracked. New technologies may improve productivity for organisations, but it also reduces personal autonomy and privacy for workers.

Sometimes it is thought that these changes in work apply only to the developed economies. However, the emergence of a global economy alters economic structures even in less-developed societies whose economies do not appear to be 'information' based. Low paid factory workers in Indonesia, for instance, may be industrial workers, but that industrial work is only possible because of a global information economy that makes it possible for tasks and production to be co-ordinated on a global scale. There would be no employment if that factory were not part of an integrated global system. A multinational, headquartered in Europe or North America, can ensure that goods are produced in remote locations according to an appropriate design, using raw materials that may be sourced from another part of the world, and whose output is then delivered to markets elsewhere in the world. Similarly, there are single crop farmers in Africa, whose labour is unconnected with technology or even mass-production, but whose output is sold in Europe or the United States because of a global distribution process, linked and co-ordinated by information technology. Work is everywhere altered by the impact of new technology.[4]

4.5 Information as a commodity

New technologies have transformed existing industries, but have also created a new information industry: the production, distribution and consumption of information 'products'. To some extent, this is a subset of the service industry. For instance, producing, distributing and buying music would seem to be part of the service industry. Similarly, reports on purchasing trends in South America, publishing books, applying for housing grants all seem to be part of a 'service' industry. However, these are commodities whose value derives from interpretation of data or from original intellectual activity, and the value of the information is greater than the value of the physical objects on which the information is inscribed (see Stehr 1994: 160–202 for a discussion of 'knowledge' work).

The difference between the cost of the artefact and the value of the content is crucial to the information industry. Most of the cost of a car derives from the cost of producing and distributing the actual object, with a small amount from the creativity of the design process – the cost of the raw materials, the cost of the factory where the materials are assembled, the cost of the labour to do the assembly, the cost of transporting the car to the seller. The value of a computer programme, however, does not derive from the cost of the CD-ROM on which the programme is written or the cost of writing the programme to the CD or the cost of transporting the CD to the consumer. The 'added-value' derives from the intellectual creativity that led to the computer programme. How is the value of an information commodity to be determined? The simple answer is that in a free market, value is determined in the same way for all commodities: however much it is worth to the buyer. That worth is determined through a complex process. For any commodity, price is related to the cost of producing the commodity. The purchase price must be higher than the production cost, and the difference is the profit for the manufacturer. But, if the profit margin is very great, others will start producing the same commodity to make similar profits. Competition amongst producers will bring the price down to the point where the cost is slightly above the cost of production and distribution.

However, this process differs somewhat for information-based products as compared with other commodities. Whether information arrives on a floppy disk, a CD-ROM or via the Internet, the cost of embedding the information onto a physical artefact or transferring via electronic networks is minimal. Writing information onto a CD costs a few cents, and the cost of the CD itself is a few cents; the cost of sending an email message over the Internet is equally minimal. Thus, the cost of distributing information is slight and is being reduced all the time. The major cost of information is the intellectual work involved in producing information. In the software industry, this is the cost of paying people who can develop software. Data analysis requires people who have the expertise and experience to transform raw data into knowledge. The creation of music or video involves the creative efforts of musicians, based on their abilities as well as training. All information industries have the cost of intellectual capital – rewarding individuals whose training, experience and abilities enable them to produce such output. How does one determine the value of human inventiveness, intuition, experience or ingenuity? When an architect produces an innovative design in two hours, how does he or she add in the cost of the years of previous experience, including education, which led to the production of an idea that others could not have produced in two years? When a business expert produces a plan to save a firm from bankruptcy, how does one calculate the cost of the previous experience that provided ideas? It is very difficult to determine the input costs for the production of information. This leaves the market answer as the only mechanism for determining consumer price: if the purchase price is significantly higher than the production cost, then others will deliver designs, lectures or business solutions at a lower price.

It is the determination of the market value of information that causes problems. For other commodities, competition ensures that roughly similar products will cost about the same. However, similar value is harder to determine when dealing with information commodities. Some private secondary schools may be more expensive than other schools, but parents do not necessarily assume that all private secondary schools are equally good and so may not opt for the cheapest. A school with a very good reputation has a monopoly position; even if another school can offer lectures for a reduced price, students (and employers) may value the other school more highly and so the school can continue to charge high fees. A 'good' architect (one with a good reputation) can charge more for his or her services, as a good painter can for his/her pictures. The value of the commodity is increased because the information cannot be obtained through any other means. That is to say, the value of information derives from a monopoly position. One buys software from Microsoft because a similar commodity is not otherwise available.[5] This is one reason why intellectual property has become such an important issue; without such 'property rights', information producers could not maintain control over their product.

How can one determine which part of the commodity is based on unique intellectual effort by the company, and which part derives from commonly held ideas? The modern database depends on the concept of a 'record', which is composed of 'fields'. So in a library database, each book is represented as a record, and each record will list fields such as 'author', 'book title', 'date of publication'. Is this an original creation, which can be owned as intellectual property, or simply the adaptation of commonly held knowledge? Judges, courts and solicitors have to decide rights on issues that are both technical and complex. The 1998 US antitrust case involving Microsoft illustrates this well. For instance, Microsoft argued that the Internet Explorer browser software incorporated into its Windows 98 operating system is not a separate software 'application' but rather a 'set of technologies' that are thoroughly integrated in Windows. The government, on the other hand, noted that the Microsoft dictionary did indeed define a browser as an 'application' (Chandrasekaran 1998b). In the same case, a 'computer science professor working for the government and a lawyer for Microsoft Corp. spent much of yesterday at the company's antitrust trial sparring over an arcane question: What is a personal computer operating system?' (Chandrasekaran 1998a) These are the kinds of esoteric issues about which courts are now expected to make determinations.

The gap between purchase price and production cost is further complicated by the minimal cost of duplicating electronic information. For a car, the major cost in production is the cost of the materials and the cost of the machinery and labour in the production process. A second car that is based on the same design is still going to entail similar production costs, as new raw materials and new labour have to be used to produce a new product. In contrast, information, once produced, can be reused. The cost of a second copy of a computer programme is significantly less than the cost of producing the first copy of the programme since, once the programme has been created, there is little further significant cost in making and selling additional copies. The value of the commodity may remain the same (that is, the purchaser still needs the computer programme), even though the production cost plummets. So how does one price multiple copies of information? Again, monopoly affects price. Laws of copyright have been developed to assert that when an individuals buys information he or she is actually buying only the right to use information in specific ways. Most commodities can have only one owner, and a sale transfers the commodity from one person to another. Ownership of artefacts such as houses, cars and radios transfers when the sale takes place, and the new owner has the right to use the house, car or radio in any way he or she wants (as long as that use does not infringe the rights of others). When one buys an information commodity, one often only gains 'use' rights rather than ownership rights. The seller of the information still determines the use of it – whether the software can be used on multiple computers, whether copies can be made and distributed to others.

4.6 An information economy or an information society?

It is clear that information has become a commodity, although sufficiently different from previous commodities that problems are still being resolved in both the economic and the legal arenas. It is also clear that information has suffused all economic sectors, however traditional those sectors might appear to be. The impact of this transformation is all around us; it has changed the way we earn our living, it has changed the kinds of things we can buy, and the way we can buy them. It has changed the nature of work and is transforming the world into a global economy. We now live in an information economy, rather than an agricultural or industrial economy. The emergence of this information economy is largely due to new information and communications technologies. In this sense, Daniel Bell has been proven correct; new technologies are transforming economic activity. Yet not all the features of the industrial economy have disappeared. Factories still produce goods. There are still those who own economic resources, and those who sell their labour for a wage. So arguments about post-industrial versus advanced industrial continue.

This information economy has an impact on the way we live our lives, but, just as technological change does not determine particular social outcomes, neither does economic change, in and of itself, determine the nature of society. Does an information economy herald the beginning of an information society? An information society is not just the impact, however significant, of information technology and changed consumer preferences on the global economy or individual work conditions, it is one in which there has been significant social, political and cultural change. As the Finnish case shows (Castells and Himanen 2002), it is possible for societies to have similar economies but radically different societies. Economic forces influence but do not determine political structures, social relations or cultural beliefs. Economic structures are themselves affected by broader social forces, as is technology. It is not possible to determine the nature of an 'information society', or even the existence of such a society, simply by describing technological or economic changes. There must be broader changes as well, and these changes are not going to be inevitable consequences of either technological *or* economic change.

Chapter 5

An information society

What is an information society, if not a society with a high degree of technological development and investment, and with an economy in which technology controls the production, distribution and consumption of goods and services? Without minimising the importance of both the technological and economic changes that are taking place, an information society is too grand a term for an increase in the number of computers, or even the development of a new economic sector. For such changes, a technological society or information economy would be a sufficient description. Surely the information and digital revolution has had more than a technological and economic impact, and there must be social, cultural and political changes that are not simple side effects of technological and economic change. Such changes are difficult to quantify or measure, so perhaps it has been inevitable that so many discussions about the new information society provide impressive statistics about increases in computer use, or bandwidth, or economic revenues generated in the information sector, while being either vague or aspirational on social, cultural and political changes. However, technologies are only means to ends; they are not ends in themselves. Social structures and cultural beliefs exist in their own right, not as dependent features of technology and economy.

Not all societies with the same technology or the same economic structures have the same social, cultural or political structures, although this is sometimes thought to be the case. The belief that societies follow a similar line of development, as dictated by technological development, was commonly held in the early days of comparative social research. Lewis Henry Morgan proposed a tripartite division of history into savagery, barbarism and civilisation when he published *Ancient Society* in 1877 (Morgan 1877). He assumed an inevitable progress from one level of development to another, and that each level (and sublevel) could be identified by its characteristic technology. Thus, as technology developed from axe to bow and arrow to bronze spear, so also did society develop. Although this view of unilinear evolution lost credibility in the early part of the twentieth century, there continues to be debate about the determinist impact of technology (Pfaffenberger 1992). The position in this book has been that while society and culture are affected by technology, the reverse is also true, and it is most useful to see technology, economy and society as interwoven, but separate, facets of human life.

To avoid a technological or economic view of the societal impact of new technology, it would be useful to focus on the 'information' part of new information technology. An information society is a society in which information places a central role in all aspects of life, not just in economic life. This makes intuitive sense, given the importance of digital information. The computer and telecommunications revolutions have changed the entire process of information production, distribution, consumption and reproduction, leading to an increase in the amount of information, and also in the amount of that information that is in public circulation, as well as an increase in the significance of that information. That is, this is a society that produces more information than before, in which more information is communicated amongst a larger number of people, and in which that information is central for the survival of that society.

For this perspective to be more useful, 'information' should be defined and measured. We all 'know' that there has been an increase in the amount of information in modern societies – bits and pieces of facts about everything under, and over, the sun. But can this increase be quantified? One early attempt at definition has come from electronic engineering, and the need for communications engineers to improve the communication of information by improving accuracy of data transmission along specific communication channels. In 1949, Shannon and Weaver, of Bell Laboratories, published *A Mathematical Theory of Communication*, which is often credited with establishing 'information theory' (Shannon and Weaver 1949). Eventually, from this beginning, came the plethora of words which we are all used to, such as bandwidth, redundancy and signal/noise ratio. These are all ways of measuring the amount of information being transmitted. The only problem is that the information being measured is not what most people think of when they use the word 'information'. Shannon and Weaver were concerned with the most efficient way to transfer data from one source to another while avoiding distortion. They were concerned with the accuracy of the data transmission, but not with the meaning of the data. Transmitting the phrase 'The President has been assassinated' accurately is the same technical problem as transmitting the phrase 'The artichoke spoke the car'. The fact that one sentence would have enormous significance, while the other is meaningless, is irrelevant as far as a technical theory of communication is concerned. On the other hand, such a difference is vital to anyone concerned with meaning and knowledge. The work of Shannon and Weaver in information theory has been crucial in the telecommunications revolution, but it is not the basis for a definition of information, and much less so for an information society.

Daniel Bell refers to information as 'data processing in the broadest sense; the storage, retrieval, and processing of data becomes the essential resource for all economic and social exchanges' (Bell 1979: 168). This is very similar to the

Shannon and Weaver definition, since information becomes units of data that are somehow independent of sender or receiver, with interpretation playing no part in the process. The problem is that measuring the amount of data is not a measure of the significance of those data. A 1500-word essay is three times as long as a 500-word essay, but there is not necessarily three times more knowledge contained in it. Nor are data of equal significance. Are data about rainfall in Brazil as important as tomorrow's stock price for Microsoft? They might be, depending on the use to which the data are being put, and the other knowledge that the person may possess. For someone with money to invest, the stock price is important. For someone planning next year's planting, rainfall is important. It is possible to measure increases in the amount of data that exist, or even increases in the amount of data in circulation (see the discussion in Duff 2000). Measuring increases in the amount of information or the amount of information in circulation is another thing.

5.1 The information explosion

There has been a huge increase in the amount of data available in the public domain, especially in the last fifty years. Even before information became central to economic life or part of a digital revolution, it had become central to political survival. Intelligence has always been an important component of military success, but the Second World War marked the beginning of an increase in the relative importance of 'information-based' military action versus 'material-based' action. Military developments such as radar, bomb targeting devices and the atom bomb made a greater difference to the outcome of the war than numbers of soldiers, tanks or ships. Equally important had been the decryption advances that enabled Allied governments to decipher Japanese and German military transmissions. Governments learned the lesson that technology and information tipped the balance of the war and invested ever larger amounts of money in research in order to produce more advanced military resources during the Cold War. Many of the technology developments that are central to the information society, such as microcomputers and the Internet, developed because of governments' investment in research for the sake of military advantage.

The second half of the twentieth century saw an increase in the amount of money invested in research, an increase in the number of people carrying out research, and a commensurate increase in the amount of information that entered the public domain (that is, information which is neither politically nor economically sensitive and whose circulation is thus not restricted to a private group). As information production increased, the number of publication outlets also increased – more journals, more books and more conferences.

Even before the transformation to digital information, an information 'industry' developed. This is a collection of institutions and activities which produce ever more information, and put that information into circulation: the number of books published has increased, the number of journals being published has increased, and now, with the advent of electronic journals, there is little limit to the amount of information entering the public domain (Price 1963; Machlup 1962; Machlup and Leeson 1978; Machlup 1980; Gibbons, Limoges et al. 1994).

The digital revolution facilitated this information explosion. Traditionally, regardless of the amount of information that was accumulated by individuals, only a limited amount could be published. Cost limitations (in publishing, distributing, consuming), whether for books to be published, articles to appear in journals, or events to be reported on news, forced publishers and editors to restrict the amount that information that could be published. Recent advances in digital publishing of information have removed the economic restrictions on the amount of information that can be produced, distributed or accessed. The space restrictions for print or electronic media do not exist for digital media. It is almost as cheap to produce an electronic newspaper with 100 'pages' as one with 25 'pages'. It is almost as cheap to have a three-hour-long digital news broadcast as a one-hour news broadcast. It is no longer necessary to decide what goes in and what is left out; it can all be published.

The technologies of traditional mass media publishing and broadcasting are expensive (large printing presses, television studios and broadcast facilities) and this effectively excludes most people from being able to produce and distribute their own television programmes or newspapers. The reduced cost of digital production and distribution technologies, including desktop publishing, inexpensive video cameras, online data storage, digitalisation of voice and visual images has increased the number of information producers. People and organisations previously excluded from information production can now become the digital equivalent of a newspaper publisher or a television broadcaster. They reach audiences they could not previously have been reached, and the variety and amount of information available are now breathtaking. People's appetite for information has increased, and the increased information output has found a ready audience.

If the amount of information circulated in the public domain has increased, so also have the number of people and factions who can now make information and opinion widely available. Significantly, this increased amount of published information is not mediated. Previously, information might be excluded from publication because it challenged the status quo of vested interests and organisations. Such interests and organisations no longer control access or 'interpret' the information produced by others. Instead of a summary prepared by a journalist or someone who might have a vested interested and biased perspective, people can have direct access and make their own evaluations.

This increase in information producers has tended to blur the distinction between one-to-many mass media and one-to-one private communication. There is now a communications infrastructure that supports many-to-many communication: a sharing of information previously unthinkable. Anyone can become both producer and consumer of information, and so a global community can emerge in which information becomes free for individuals to create, access and circulate. However, the information revolution has also increased the amount of private communications between individuals. Private communication, as well as mass media, suffered from the limitation of having to be transported physically. Letters were inexpensive but slow. Phone conversations were rapid but expensive. Face-to-face meetings involved expensive travel arrangements. Electronic mail, mobile phones, answering machines, digital photos, web cams are just some examples of technologies that permit inexpensive and rapid contact between friends and relations, regardless of the physical distance that may separate them.

Thus, the information explosion results from a number of factors. There has been an increase in the commercial as well as the intellectual benefits of information, so there is more funding for research, and more people paid to do research. There is actually more information now being created and/or discovered. In addition, information that would previously have remained unpublished now circulates in the public domain, as it is now less expensive to publish information. Information that was previously private or circulated only amongst a restricted audience is now published for mass consumption. The technology to enable storage of a large amount of information has decreased in cost. The amount of information stored for personal use has increased. Finally, the number of information consumers has increased. The result has been a staggering increase in amounts of information. The School of Information Management and Systems, University of California at Berkeley, has run a project to calculate how much 'information' is produced and/or distributed annually.[1] Their current estimate is that 'print, film, magnetic, and optical storage media produced about five exabytes of new information in 2002'. An exabyte is one billion billion, or a billion gigabytes, and the project calculates that 'five exabytes of information is equivalent in size to the information contained in half a million new libraries the size of the Library of Congress print collections'. The information revolution has led to an explosion of hardly credible proportions.

5.2 Authoritative information

This increase in the amount of information being published may have positive elements, but it has reduced the accuracy of information. Previously, published information was public information that was socially sanctified or verified, and this was underpinned by the threat of legal action. Production technology was expensive, it could not easily be moved, and therefore was a permanent 'hostage to fortune'. An inaccurate or libellous newspaper report or television/radio broadcast could lead to a court case, and it was always possible to enforce court orders – the printing press could be confiscated, or the television station could be closed down or lose its licence. Any such sanction represented a significant financial loss to the organisation, and if broadcasts or printing required state consent, then future publication could be prohibited. Even if such consent were unnecessary, there would still be a lengthy period of time before new equipment could be obtained. Therefore, information had to be validated before it was broadcast, and the public assumption was that if it was presented then the information must be true. In societies with a free press, citizens trust organisations to provide accurate information. After all, if the information is false or inaccurate, the consequences would be immediate.[2]

While mass media, whether books, radio, television or newspapers constituted 'official' or 'authoritative' information, informal information such as gossip or mimeographed handouts was not subject to the same regulation and thus was not as trustworthy. If the information was wrong, there was no one who could be held accountable, and no expensive equipment that could be seized. Luckily, such non-authoritative information was easy to identify. It was easy to tell the difference between gossip and a news broadcast, between a book and a mimeographed handout; good quality output required expensive equipment, which meant it was more likely to be 'authoritative'.

The increase in the amount of information produced, resulting from an increase in the diversity of groups and individuals that are able to produce it, has led to a greater diversity in information content. Information about genetically modified foods, nuclear energy, illegal labour practices and corrupt political practices are now available when previously it would not, or could not, have been published. In cases where the state controls and distorts mass media news, alternative publishing provides an important, and potentially subversive, source of information. Even in countries that have a free press, there may be self-imposed restrictions limiting what appears in traditional publications and broadcasts. In the United States, it was an alternative Internet publisher, the Drudge Report, who broke the Lewinsky case (Shapiro 1999: 41). The information was previously known, but traditional publishers did not publish it, because they thought it was not appropriate, or were afraid of libel, or else just afraid of offending President Clinton. On the other hand, traditional publishers have also failed to publish fervent and convincing reports of alien invasions, links between the John F. Kennedy assassination and the CIA, and other dubious revelations.

With new technologies, maintaining publishing accuracy by the threat of legal action is no longer effective. It may still be illegal to publish libellous or inaccurate information, but it is difficult to enforce these laws when the content is digital. The publisher may be outside the government's jurisdiction. It is now possible for Irish people to read electronic publications that have originated elsewhere in the world, and it is difficult to prevent transfer across national boundaries and hard to enforce legislation in all jurisdictions. As the European Union, the United States and other jurisdictions move to harmonise regulations, jurisdictions remain with different laws. Even if the published material contravenes local laws, the equipment is inexpensive and easily moveable, so it may be difficult to find the equipment and it is easy to replace any equipment that is impounded.

With new digital technologies, anyone can sound like an official radio broadcast, and mimeographed personal output can have the same authentic look as a book from a well-known publisher. Appearance, once useful evidence of authenticity, is no longer reliable. Information produced by individuals can have the same professional look as authoritative information produced by organisations; information from unofficial sources has the same look as authoritative sources. The distinction between formal versus informal, authoritative versus personal, has now become blurred and the burden for determining the authenticity of information has now shifted from publisher to consumer. New criteria for evaluating the authenticity of information have to emerge.

5.3 Relevant information

Traditional publishing not only provided authoritative information, it also provided relevant information. News reports have to fit into a finite time slot, and newspapers have a finite length. Editors have to decide which events are 'news', and which news stories are relevant and significant enough to include in a 30-minute slot or 20-page newspaper. Such decisions would not matter if information were objective fact. However, the process of selecting information requires subjective judgements. Publishers and editors judge the relevance of one bit of information over another. They learn a set of rules as to what constitutes relevant and accurate facts, and these rules constitute their 'professional' judgement.[3] Publishers and editors often claim that, as professionals, they make decisions without regard to vested interests; their decisions are based on what they believe their readers or viewers would consider news, or what they think readers or viewers should know. By default, information consumers leave decisions regarding what is or is not 'significant' information to the editors and publishers, perhaps choosing those newspapers or broadcasters that are most likely to mirror their own concerns.

The same process applies to scientific publishing. Only a portion of scientific research information can be published, and editors of journals look to referees who try to make judgements based on academic merit. But this process also has a social dimension. In practice, the language and terminology have to be in accord with that particular academic subject, the topic has to be acceptable, and it is easier to publish if the person has the right academic credentials (the appropriate degree from the appropriate university, a job in the appropriate institution). Articles written using a different style, or by individuals without recognised qualifications, or on a topic not considered 'important' will not be published, even if the actual information content of the article has merit. In effect, there are rules as to what articles are relevant and accurate, with editors and publishers making the decisions and acting as gatekeepers.[4]

Readers avoided being overwhelmed by too much information because editors and publishers sifted through the available information and decided what information was important or relevant enough to broadcast or print. There are no longer the same restrictions on the amount of information that can be published; now, as long as an author finds the information worth a small effort, it can be made available to the world. This makes information available that was previously excluded from publication. But with no editors to 'sift the wheat from the chaff', how can people find relevant information without drowning in the sea of irrelevant information?

There are a number of strategies by which relevant information can be retrieved; perhaps the two most common are word searches and classification systems. Book libraries provide an example of both. Libraries depend on a classification system to group texts together into headings and subheadings. Thus someone searching for information on a topic can find a number of books all grouped together, within the general field of geography, the more specific field of Europe, and yet more specific field of Ireland. On the other hand, if a person knows the author or title, they can find all the occurrences of that title or author, and then choose the book that they are interested in. Books themselves use the same dual system of indexes (text search) and tables of contents (classification).

Information retrieval acquires a new significance in the digital age, for both individuals and organisations. Previously, individuals possessed a small enough amount of information that it could be organised on an ad-hoc basis, although it might occasionally mean being unable to find the particular letter, book or notice that was put somewhere in a makeshift office. Now, each individual has far more information at his or her fingertips, much of it in digital format, and that information tends to be more important. As the number of computer files multiplies, problems of classifying information so that it can be found later are becoming insurmountable. Most people do not classify files into categories; all files are in the same directory. Then people have the difficulty of finding names for files that would enable them to identify the file in the future. All of us have to create, or learn, classification systems that help us organise our own personal and disparate items of information.

Information retrieval has always been an important issue for organisations: items are put into folders, and folders are part of a larger filing system. Filing cabinets are, after all, one of the icons of office life. There are often strict controls as to who can file away information, and where it can be filed, so that relevant information can be found at a future date. To ensure that everyone applies the same rules, individuals have to learn the rules of the particular organisation (Suchman 1996). However, organisations are finding that their filing systems are unable to cope with increases in the amount of information, multiple formats for information, and multiple locations for information storage, and these problems pale by comparison with the problem of locating external information, which may be published in a book, journal, video, CD-ROM, or, more frequently, on an Internet website. Classification systems developed for text libraries (whether Dewey Decimal or Library of Congress) are not satisfactory for multi-media digital information. In any event, there is no single organisation that has the resources to classify all information on the Internet. For that reason, the Internet has often been compared to a library, in terms of the richness and variety of information available, but a library in which the books have been scattered at random all over the floor, and authors just toss any new books onto the floor at random.

What are the options, if one is searching for information? One solution is to pay someone else to find and classify information. Professional companies obtain information, first validating it, then classifying it, interpreting it, and then selling the results on to others. The process is costly since it requires special experience and training to be able to find, evaluate and interpret information. Another solution is to visit sites provided by an individual or organisation that does all this at no charge. Many people depend on 'portals', which are sites that specialise in collecting information on a particular topic, such as health or sports or computers. They may do this in order to advertise themselves, or they may do it as part of their function. Many government sites, for example, do this as a public service. Another option is to depend on the electronic publication sites of traditional publishers – newspapers and television stations often have a website in which they provide relevant and accurate information. These strategies require that individuals once more become dependent on others who determine if information is accurate and relevant.

Are there any means of finding information that keeps individuals in control? With the increased amount of information on the World Wide Web has come a range of computer programmes to help find information. Yahoo, AltaVista and Google are just some examples of recent efforts to help individuals find digital information. Computer searches for information suffer from a number of problems. It requires training to learn how to use a search programme effectively, yet most people simply type in a few keywords to find information. People commonly get either too many or too few responses, and it is virtually impossible to decide which is the most relevant response. The computer programmes are designed to get around these problems, perhaps by ranking information by its popularity, or number of links from other sites. Others can detect the pattern of choices and interests and provide only 'relevant' material. Unfortunately, it is still difficult for such computer programmes to evaluate accuracy.

Given the dubious quality of the information that is often available, the difficulty in finding information, and the uncertain significance of information, why should we be impressed by a growth in the amount of information in modern society? Why should being able to find out the temperature in Tokyo or the annual rainfall in Brazil mark a transformation in our lives important enough to warrant such a grandiose term as 'information society'. On the contrary, we are drowning in information, some of it accurate but irrelevant, some of it both inaccurate and irrelevant, some of it accurate and relevant but impossible to find, and, only occasionally, information that is accurate, relevant and obtainable. An increase in such a mix of information is hardly epoch making. However, such measures miss an important point: the digital revolution is not a growth in the amount of information, but a growth in the social significance of information.

5.4 A knowledge intensive society

For many people, one indication of life in an information intensive society has been information overload: too much information, but no way to decide which of it is actually important or accurate. This overload is both passive and active. Anyone searching for information on the Internet has suffered passive information overload: finding too much information, and no way to shift through it all to find the information that is actually useful. Equally problematic is active information overload: too many email messages, text messages, messages on answering machines, letters in the post. The overload comes from an anxiety that some of the messages may be important, so they cannot all be ignored or deleted. Instead, message after message has to be read, just in case one of them is crucial. We are all aware that we live in an information intensive society, where one ignores potentially important information at one's peril. In short, we live in a society in which we are dependent on information.

The amount of information that circulates in the public domain has increased, but the amount of information any single individual possesses has not necessarily increased. Cars may be more complicated and so there is more to know about how they work, and the same can be said for nuclear energy or treating stomach cancer, but individuals do not possess all this information, do not need to know about it, and may even find it difficult to learn about it. This increased knowledge is held by society collectively; some of it is an external memory and stored in devices such as books and computers, while other information is distributed amongst members of society who have specialised knowledge. We are dependent on knowledge that is distributed throughout society, yet we have access to a small amount of that knowledge. Indeed, we only need access to a small amount of it; the rest we either do not need (depending on experts who do know) or can find as necessary.

It is our increased dependence on distributed information that signifies a knowledge intensive society. New technologies of information have altered control and communication in society, exemplified by changes in the spatial boundaries of control and the amount of control that can be exercised within that space. Economic exchange provides an example of this. The earliest form of economic exchange was a barter system in which value was negotiated at the time of the exchange. It was necessary because there was no system in place by which the value for a good or service was agreed in advance. Once a value is attached to goods and services, along with a common payment system, then the circulation of goods and services becomes easier since there is no need to negotiate, *ab initio*, every transaction. The boundaries of an economy are determined by the boundaries of common exchange, and the limits of the information system define the limits of common exchange. One of the clearest results of the information technology revolution has been the expansion of the spatial limits of information exchange, so that goods, services, capital, and labour all circulate within a global economy.

The spatial expansion of information has also enabled the co-ordination of diverse (both technically and spatially) economic activities. Previously, all parts of the manufacturing process were in one location because this was the only way to ensure that the entire process functioned smoothly. New information technologies permit the co-ordination of labour and raw materials; it is possible to ensure smooth manufacturing even if different segments of the production process are distributed throughout different parts of a region, nation or world. Poor communication used to require the delegation of responsibility for economic decisions and limited the amount of co-ordinated action that was possible. New technologies enable the co-ordination of even the most minor economic activities in a global system.

The political consequences of technology have been as important as the economic consequences. An increase in the spatial boundaries of states owing to new technologies of information control and co-ordination have also enabled states to exercise greater control over the citizens that live within those boundaries. We live in complex societies that depend on the routine accumulation and co-ordination of vast amounts of information.[5] The continued growth of the states and the emergence of super-states (e.g. the European Union) are dependent on the improved information processing and storage that is now possible.

The amount of information that individuals disclose about themselves to governments and private organisations in their everyday lives has also increased. Individuals leave 'footprints' every time they have a transaction. If they buy something, information is recorded about what is bought, when and where it is bought, and how it is paid for. Storekeepers routinely know how much stock they have, and what the general trends in sales are. Companies can track sales, according to region, type of product and distribution point, and use all this information to plan the future of the company. Governments collect a great deal of information about individuals which they analyse in order to make policy decisions about future trends, as well as to monitor access to government entitlements. The amount of information on individuals stored in records – mobile phone records, social welfare records, credit card and bank records, telephone records, education records – has led to a loss of privacy for individuals and a growth in the apparatus of the state.

Equally significant has been a change in the distribution of information. It was a characteristic of early, non-state societies that most information was equally distributed and that everyone shared a common stock of skills and knowledge.[6] The development of writing and then printing, as we have seen, led to inequalities in access to information. Those wealthy enough to have learned to read had access to information that was not available to poorer, non-literate citizens. Even amongst the literate, there would be some who were more 'learned' than others. That differentiation has now taken on a different character. Society is now so complex that we are always dependent on other people's knowledge. Furthermore, that knowledge is differentiated. That is, different bits of knowledge are held by different people. Sometimes this expertise can be purchased as part of market exchange; thus, part of the cost of repairing a dishwasher is the cost of the knowledge and experience held by the person repairing the dishwasher. Similarly, we purchase the expertise of lawyers and doctors. Not all information is of equal value and some expertise costs more to obtain than others. Furthermore, the price of that expertise may vary from one expert to another. Not only might a surgeon charge more for his or her services than an insurance agent, but the surgeon might also not charge a friend the same fee as other patients. Even in a free market not all information is available to all. The control of knowledge becomes both a political and an economic issue, out of which develops the slogan 'information is power'. Even in war, success is not determined by the amount of labour (soldiers) or the amount of capital (number of missiles), but knowledge (smart bombs, cruise missiles). Contemporary societies are both information intensive and information dependent.

The public is increasingly aware that information is used to influence and control people. There was a time when scientific evidence was accepted as evidence that a proposal or plan was beneficial, even if individual members of the public were uneasy. That time is now passed. When 'scientific evidence' is used to justify incineration for waste disposal, nuclear reprocessing or genetically modified organisms, people who oppose such projects do not simply reject particular facts, they reject the entire premise of scientific argument. They do not trust 'scientific evidence', because they see such information as being used to advance the interests of particular groups or organisations instead of being objective and authoritative descriptions of reality. They suspect that they are receiving only the information that others want them to receive, and not the whole story. People will say clearly and simply that they know something to be true, and that they do not care about contrary evidence or arguments. Information has become a contested currency, illustrating again its centrality in contemporary society.

5.5 A connected society

Not only has the amount of information circulating in society increased, but so too has the level of communication amongst individuals. Information technologies have facilitated greater communication amongst individuals. If one catchphrase of the information revolution is 'information is power', another is that new technologies have led to the 'death of distance'. Geography no longer restricts communication since digital information can be communicated from any location to any other location, virtually instantly and inexpensively. Access to services does not depend on location. Service providers can be situated anywhere and accessible by citizens via email, the World Wide Web, or even reduced rate telephone calls. This may include government services (social security, driver's licences and other government services which can be located anywhere), but also commercial services. One can access banks or shops via phone, email or the World Wide Web. These changes can benefit individuals, since consumers have access to goods and services without physically going to a store, and, by shopping around, can obtain goods at competitive prices. It can also mean significant improvement in services; a government may not be able to afford to have a radiologist in every town, but can have an inexpensive scanner and radiographer in every town, with x-rays sent via broadband to a specialist in centrally located hospital. Such distributed service provision can also result in cost savings: labour can be cheaper in a rural area than in the urban area where the service previously had to be centralised.

Do people use this global information system? The evidence suggests that, by and large, people remain interested in local information: when scanning newspapers, radio, and television news reports, it is the local content that interests people. Whether they access that information via traditional media outlets or via the Internet is a matter of convenience and cost. However, people reach out beyond their locality for a number of reasons. Individuals with specific interests make the effort of finding specialised information. For instance, if they are expatriates, they can access information from their home. If they have a specific leisure interest, they can look for specific information on football, or soap operas, or music television. If they have specific ideological interests, they can look for information on nuclear power, or a religious sect. If they suffer from a particular medical condition, they will look for specialised information sources. For such people, the limitations previously imposed by the geographical restrictions of mass media enforced a false consensus. Individuals who felt 'out of synch' with their neighbours could do nothing about it. The advent of new technology has allowed a greater diversity to be expressed in communities, a diversity that may have been present in latent form, but previously suppressed.

Our sense of the world has been transformed by increased connectivity. An obvious example is our ability to obtain news about events anywhere in the world, virtually as the event happens. The impact on individuals of the destruction of the Twin Towers in New York on 11 September 2001 was heightened by the live broadcast of the second plane crashing into the skyscraper, and the continual broadcast of events, including the collapse of the two buildings. The result was a simultaneous global experience – a shared catastrophe. Equally poignant, however, was the ability of individuals to communicate with family, anywhere in the world, as the tragedy was unfolding, even to the point of victims being able to talk to spouses on mobile phones as well as leave messages for their loved ones on answering machines.

With no geographical boundaries on information flows, individuals can access television broadcasts, national (and local) radio broadcasts, and national newspapers via digital technology from anywhere in the world. The same programmes available in Ireland are often available to anyone else in the world, as long as they have Internet access. It also follows that people in the same locality no longer have to depend on the same information sources. In Ireland, one person may be listening to an Irish radio station, broadcast via transmitter, while their next-door neighbour could be listening to a radio station in Nebraska via the Internet, and the third could be listening to German radio received via a satellite dish. Indeed, the neighbour on the other side could be watching a television station in Japan, which broadcasts video as well as audio via the Internet, while another neighbour watches Italian news via a cable connection.

An increased level of personal communication is another aspect of contemporary life: inexpensive telephone calls, email messages, even live video from one member of a family to another, anywhere in the world; chat rooms on a vast variety of topics, with members contributing, in real time, from all over the world; information on weather forecasts in cities halfway across the world; plane arrival times at a local airport; or movie times for the local cinema. We now live in a world where people need never be out of touch, and, increasingly, feel anxious or isolated if they are out of contact with friends and family. This instant access, at an electronic level, often contrasts with isolation at the physical level of local neighbours, creating an imbalance between two spheres of communication.

5.6 An information society

At this point, it is clearer why the digital revolution is more than an information economy or a technological society. Bandwidth, or numbers of computers, or numbers of automated teller machines in a society do not define an information society. We now live in a transformed economic system, which has seen the emergence of a new economic sector of information, as well as information becoming the determinant of profitable activity in most other sectors. But, though we now work in different ways, in different locations, and often for companies that are structured radically differently, this describes only an economic transformation, not a social or political one. Slogans such as a 'wired society', the 'death of distance' or 'information is power', illustrate how new technologies are also transforming social and cultural life. It is society – the relationships that people have with each other – that is being transformed, because more information is being created, with more of that created information becoming publicly available, a larger percentage of the population having access to that public information, and greater levels of communication amongst greater numbers of people. This abundance of information permits a greater diversity of views to be expressed, as well as more choice on the part of individuals as to what information they access. An information society is a knowledge intensive society, in that our everyday lives are structured by the information we access or the information about us that other people and organisations use. This information interdependence parallels an economic interdependence. This dependence is so great that, if the infrastructure supporting such information exchange fails, there would be catastrophe (as illustrated by concerns over the 'millennium bug'). Up to this point, no society could be said to be dependent on information to survive; but without constant information flows, our modern society would simply collapse.

What are the detailed political, social and cultural issues that arise in this digital revolution? How will politics change? How will domestic life change? Will there be fundamental changes in the way we relate to each other, and the way we interact with the world at large? Optimistic or utopian commentators often paint a picture of the future that seems revolutionary: we will be living in comfortable homes, interacting with the external world via technology, with the implication that all of our needs and wants will be catered for. Pessimists describe a world in which new technologies are used as agents of control and manipulation, in which individuals' needs and wants are redirected rather than satisfied. The actual outcomes of new technologies will depend on the policies which individuals and governments put into place.

Chapter 6

Regulation of information

6.1 Introduction

Some of the changes arising from the information revolution resulted from a clear political agenda. Up to the 1970s, computing was the preserve of the wealthy organisations that could afford huge mainframe computers. The microcomputer was the technology that would enable individuals to afford computers. The agenda was to provide affordable computers for those who were excluded or marginalised; this would improve their position and counter the power of the wealthy and powerful of the industrial society. It has been an idealistic aim of those promoting new technologies to broaden democratic participation, enhance accountability of elected politicians and civil servants, and to promote a committed and informed citizenry. Technology, it was hoped, would redress the imbalances between those who have traditionally held power and those who have been excluded from power.

However, the political consequences of the information revolution have been greater than redressing a power imbalance between individuals and organisations. Information rich societies are also more complex societies and new technologies are integral to the process of administering them. In addition, government has always had a spatial dimension: the state has been a geographically bounded entity, with sovereignty defined as control over what happens within that entity and control over movement in and out of that entity. New technologies are blurring this boundary, creating new problems for control over sovereignty. Information is a new sort of commodity, and electronic commerce is a different sort of exchange; new laws and regulations are required.

The political consequences of the information revolution fall into three categories: participation, administration and regulation. Generally, participation is the process by which citizens and their representatives make policy decisions regarding the allocation of scarce public resources. By definition, it is an arena for conflict and opposition, as varying policy options are articulated by different members of society and a consensus must emerge. Administration is the means by which those policy decisions are carried out, usually by a government bureaucracy composed of civil servants. Regulation is the administration of government policies that affect the activities of individuals and organisations outside the direct control of the state. In all three areas, government activities are changing because of new technology.

6.2 Electronic government

Organisations process and store information, and information technologies have made improvements possible in both efficiency and reliability. This is as true of government bureaucracies as it is of any other organisations, except that everyone is affected by government bureaucracy, including residents in the country (citizen or not), citizens who may not be resident in the country, organisations both within and outside the state's jurisdiction, and other countries. New technologies permit changes in administration of policy and the provision of services, which have been necessary owing to the changing nature of government. The increased sophistication of information technology over the last fifty years has been matched by increased state intervention in providing benefits to, and regulating the activities of, individuals and organisations. The modern state is expected to provide a wide range of benefits not previously available to citizens, and to regulate a much wider range of activities than before. Investment in information technology has enabled the state to provide more services and in a more sophisticated way.[1]

Bureaucracies traditionally dealt with paper-based information. Indeed, a common description of bureaucrats would be 'paper pushers'. With increases in the amount of information processed by governments, paper-based records have become an impediment to efficiency. Paper records have to go from person to person, they have to be stored, they have to be accessed from storage if something has to be checked (which means indexes and searches), and new information has to be added to existing files. Errors creep in when data are input or duplicated, delays result as the paper files move from person to person and can get lost, and storage costs and access costs for such records can increase almost out of control. All organisations have found that the transformation from paper-based records to electronic records speeds the flow of information, improves the accuracy and efficiency of adding new information to existing records, reduces the cost of storing records, increases reliable and speedy access to records, and permits sophisticated analysis of existing data. Such benefits have been imperative for government bureaucracies drowning in paper, and, even without any necessary decrease in labour costs, they have been sufficient to drive most governments to move towards electronic information systems.

Many governments are implementing policies whose objectives are to improve the delivery of services by the use of information technology. In the United Kingdom there have been a series of government initiatives (Robins and Webster 1999: 64–5), including the Information Society Initiative, and the National Strategy for Local e-Government. More recently, the Office of the e-Envoy (www.e-envoy.gov.uk), was set up with the strategy to 'get the UK online, to ensure that the country, its citizens and its businesses derive maximum benefit from the knowledge economy'. Two of its objectives were to deliver electronically, and in a customer-focused way, all government services by 2005 and to co-ordinate the UK government's e-agenda across different departments. The Irish government's eGovernment strategy, most recently articulated in October 2003 (Information Society Commission 2003), restates OECD objectives of 'enabling better policy outcomes, higher quality services, greater engagement with citizens' (Organisation for Economic Co-operation and Development 2003a), and aims for 'more efficient administrative procedures, delivery of higher quality services and provision of better policy outcomes'. This policy builds on an earlier goal of having all public services that are capable of electronic delivery available online through a single point of contact (Information Society Commission 2002). These require restructuring of administrative processes, especially to provide information in terms of citizens' needs rather than traditional government structures. These aims are shared by many governments throughout the world. Governments are implementing these policies to achieve the same efficiency and productivity gains that other organisations have achieved from technology.

Once information is in digital form, new possibilities develop. For instance, new technologies enable changes in the location of administrative services. With a freephone number, it is possible for a citizen to contact a government department, even if that department is located in a regional or peripheral area such as Donegal. It is possible to transfer relevant electronic files between departments or department sections, whether those sections are across the corridor or across the country. This permits a decentralisation of administrative services that would not previously have been possible, with a single government department composed of sections that are located in different parts of the country, but still available to all other departments as well as the general public. A civil servant can access an electronic record, in response to a phone call, even if the phone call originates in Dublin, the civil servant is in Donegal, and the electronic record is stored in Dundalk.

Digital information is inexpensive to store, inexpensive and easy to transfer from one file to another, and from one department to another. Matching a growth in the amount of information on individuals and activities has also been an increase in the extent to which this information can be shared amongst all civil servants. It also becomes possible to analyse in minute detail this increasing amount of information. A more sophisticated analysis of data becomes possible, enabling more complex administrative procedures and more sophisticated policy decisions. The use of smart cards to provide payments for citizens (social welfare, health, unemployment) not only reduces labour costs, it also provides more independence and autonomy for those dependent on social welfare. Individuals can claim benefits without having to queue up in an office. This can be a double-edged benefit, however. As the process becomes increasingly complex, the entire system becomes dependent on rules that only computer programmers can understand. Automated information systems are less flexible than human-driven systems, and the benefits of independence and autonomy might ring hollow to those who fall foul of the computer system, and would prefer a human being who can override a rigid information system.

Governments are organisations with special security requirements. In order to gain efficiency benefits from providing services electronically, government departments must be able to verify the identity of citizens. In the past, civil servants verified the authenticity of a citizen using signatures or photographs; the person in front of the civil servant was, indeed, the person in the file. These mechanisms required the citizen to be physically present so that their identity could be authenticated. In order to take advantage of new technologies, there must be procedures to verify an individual's identity without the person being physically present. In the commercial world, pin numbers, birth date and mother's maiden name are often sufficiently secure validation, and organisations provide information about bank accounts or purchases based on such authentication. However, these are relatively easy to circumvent. Individuals can gain sufficient information to circumvent validation procedures and impersonate someone else in order to obtain cash or credit. Cases of 'identity theft' are increasing. Government interactions with citizens require a more secure mechanism of electronic identification to provide authentication. In Ireland, it is proposed that citizens have a single identification number (Personal Public Service Number), and for all relevant information about a citizen to be recorded in a single data location – a central data vault or e-government broker (Information Society Commission 2003). Citizens or government agencies will be able to access the relevant portions of information about the citizen, thus reducing the need for obtaining the same information over and over again. However, until the technology enables secure validation of identity, many of the benefits of electronic government will be restricted to non-sensitive information that would be available to anyone. This would include finding out about general entitlements, but would exclude information about personal circumstances.

The information revolution has led to an increased amount of information that can be obtained and stored. When this is information about individuals and is held by governments (as well as other organisations), there is a reasonable concern to know how such information is used and how personal privacy is protected. 'One-stop shops' enable citizens to access the entire range of services via one contact point, but such shops require more information about citizens to be shared amongst ever more civil servants and government departments. What are the limits of a government's rights to collect, share and then use information about its citizens? What rights does a citizen have, with regard to all this information that governments now collect?

In general, these issues fall into two general areas: data protection legislation and freedom of information legislation. Data protection legislation deals with information collected about individuals by any organisations, while freedom of information deals with information held on a variety of subjects (including personal information) by governments. The two overlap, however, since information held about individuals by governments may be subject to either or both.

6.3 Freedom of information

Government departments routinely collect information in order to administer policy. Once a record is kept when a citizen interacts with a government department, there is no end to the amount of personal information held about citizens: social welfare information, employment histories, home and auto ownership, television ownership, medical history, credit information, and more. It may not even require collecting the information, but just 'capturing' the information as it flows through the electronic information system. The information can easily be shared amongst other government departments, and easily analysed to create a detailed picture of the individual. But is this detailed picture an accurate picture? Do citizens know what information is collected? Are there limits to the information governments can collect and what can be done with it? Should citizens be able to know what information is held about them, and be able to challenge its accuracy?

In recognition of the increased importance of information, and the damage that can follow from incorrect information, many jurisdictions have introduced legislation that gives individuals rights with regard to information held about them. Loosely termed 'Freedom of Information', these laws usually enable individuals to obtain a copy of personal information held by any government department, gives them the right to challenge the information and have it amended if the information is wrong, and obtain reasons for decisions affecting them. As the Irish Freedom of Information Act, 1997, states, such laws are intended 'to enable members of the public to obtain access, to the greatest extent possible consistent with the public interest and the right to privacy, to information in the possession of public bodies and to enable persons to have personal information relating to them in the possession of such bodies corrected' (Government of Ireland 1997). Personal information, in this context, is taken to mean 'information about an identifiable individual that would normally be known only to that individual or members of the family, or friends of the individual. It also includes information about an identifiable individual held by a public body on the understanding that it would be treated as confidential.'

Sweden passed the first 'Freedom of Information' law in 1776 (Freedom of the Press Act). It required that official documents should 'upon request immediately be made available to anyone making a request' at no charge and the Act, now part of the Constitution, decrees that 'every Swedish citizen shall have free access to official documents'. Only in recent decades have governments around the world enacted similar legislation enabling access to both personal information held about individuals themselves and information about the activities of public bodies. Over 40 countries now have comprehensive laws to facilitate access to state records; over 30 more are in the process of enacting such legislation. In Western Europe, only Germany and Switzerland lack legislation, and nearly all Central and Eastern European countries have adopted laws (Banisar 2002). Legal provision for such information does not, of course, necessarily mean actual access to information. Access and enforcement mechanisms may be weak or unenforceable, and governments often try to avoid releasing information, engaging in delays or creating obstructions (such as high fees). The move towards increasing access and openness has been reversed since the terrorist attacks of 11 September 2001; governments, sometimes with good reason and sometimes as an excuse, have restricted access to information to help safeguard the public good.

One of the best-known countries for Freedom of Information access has been the United States. The Freedom of Information Act (FOIA) was enacted in 1966, with a recent amendment in 1996 by the Electronic Freedom of Information Act, which provides access to records in electronic form. Legislation allows any person or organisation, regardless of citizenship or country of origin, to ask for records held by federal government agencies. There are nine categories of exemptions, including national security, internal agency rules, information protected by other statutes, business information, inter- and intra-agency memos, personal privacy, law-enforcement records, financial institutions and oil wells data. The Act is highly utilised: 2,235,201 FOIA requests were made to federal agencies in 2000 (Banisar 2002). A related act, the Privacy Act of 1974, allows individuals to access their personal records held by federal agencies.

The United Kingdom acted somewhat later than the United States. The UK Freedom of Information Act was approved in November 2000, and provisions of the Act that allow citizens to request information will not be enforced until 2005. In Ireland, Freedom of Information legislation, while not enacted as early as Sweden or the United States, still predates the United Kingdom legislation. The Irish government examined the strengths and weaknesses of previous FOI legislation elsewhere, and considered those issues when framing the Irish laws. The Irish Freedom of Information Act, 1997 was approved in 1998, and is intended to 'enable members of the public to obtain access, to the greatest extent possible consistent with the public interest and the right to privacy, to information in the possession of public bodies and to enable persons to have personal information relating to them in the possession of such bodies corrected' (Government of Ireland 1997; see also McDonagh 1998). Citizens are given three explicit rights:

- a legal right for each person to access information held by public bodies
- a legal right for each person to have official information relating to himself/herself amended where it is incomplete, incorrect or misleading
- a legal right to obtain reasons for decisions affecting himself/herself.

While it originally applied only to Government departments, its remit has expanded to include local government departments, semi-state bodies, and all other institutions funded by the state such as universities or hospitals. The legislation assumes that information is to be made available, unless the information falls into specified categories of permissible exclusion. Even then, applicants may appeal against any refusal and the Freedom of Information Commissioner can determine that the information release is in the public interest. As the Commissioner himself has noted:

> The Act provides that access to information is to be given 'to the greatest extent possible consistent with the public interest and the right to privacy'. Several of the exemptions in the Act . . . are themselves subject to a public interest override. In practice, this means that in the majority of cases where a request falls [*sic*] to be refused on the basis of a particular exemption, there is a requirement to consider whether that exemption should be set aside in the public interest.' (address by Information Commissioner, Freedom of Information Annual Conference, Dublin Castle, 10 October 2002, http://www.oic.gov.ie/23ca_3c2.htm)

Most Freedom of Information legislation regulates not only personal information, but also 'public interest' information. This is information that governments obtain in order to help make decisions, or information about the way in which governments make decisions. Since governments are collecting information and making decisions in the names of its citizens, it is often the case that Freedom of Information legislation enables individuals to obtain such public interest information. Making such activities accessible to citizens makes the operations of government more transparent, and is intended to increase support for the process of government. It is a relatively inexpensive means of increasing citizens' sense of participation in government, since it is often just a question of making available to citizens information that is already available in electronic format to civil servants.

Freedom of Information legislation can have far reaching effects, especially by virtue of the right of individuals and organisations to scrutinise government policy by requiring information about procedures and policies to be available to any citizen. Statistics indicate that 20 per cent of Irish FOI queries in 2001 were made by journalists (Government of Ireland 2002a), and the responses were used to query government activities. The journalists' scrutiny is as effective as the scrutiny ministers face by parliaments in parliamentary democracies. Equally important is the ability of individuals to require organisations to make explicit their decision-making processes, and the criteria used to make individual decisions. More than half of the 15,428 FOI queries received in 2001 dealt with general information, rather than personal information about the applicant. While some those queries came from journalists and business interests, at least 20 per cent 'non-personal' queries came from citizens themselves. Information released under FOI cannot, on its own, change decision-making processes or remedy incorrect decisions. However, if the process or the decision outcome contravenes existing laws, then the information can be used as the basis for legal remedy in the courts. Thus, Freedom of Information legislation not only helps protect the individual against the misuse of information by governments, it also helps individuals ensure that governments comply with their own policies in an open and transparent manner.

6.4 Data protection

There has been an immense decrease in individual privacy in recent years, and it has become impossible to avoid leaving information 'footprints' in the course of daily life. While Freedom of Information legislation enables citizens to have access to information collected by government, it does not regulate the use of information by either governments or organisations. In the past, there were pragmatic restrictions on the amount of information that could be collected, as well as restrictions on what could be done with it. Information had to be collected by hand and then written down in files. The files would have to be copied, collated and then analysed. All of this was so labour intensive that only the most important information was obtained, and most citizens retained their privacy. With the digital information revolution, information can be collected automatically as individuals shop, bank, fill in government forms, renew insurance, pay bills. This information is often entered by individuals themselves as they fill out an electronic form or participate in any electronic transaction. Every time an individual pays for items with a credit card, uses a mobile phone, writes a cheque, uses a store loyalty card, a record is created. If you have mobile phone, your movements can be tracked everywhere you go. If you travel by car, your licence plate may be monitored by video cameras. If you walk, closed circuit television may be tracking your movements and, if linked with face recognition software, could identify you personally. If you use a bus pass, it may be scanned in when verified, leaving a record of your travel. This information can easily be copied, distributed to many other parties and analysed; creating a portrait of where one travels, with whom one talks, what food one purchases, what movies one rents, what books one reads, and how often one goes to a pub have become a matter of inexpensive simplicity.

Organisations often go out of their way to collect information about individuals. For instance, when people paid for groceries with cash, there was no link between individual, store and consumer item. As people started using credit cards, there was a link between individual and store (where and when you shopped and how much you spent), but no record of items purchased. With electronic cash registers, it became possible to associate items purchased with the credit card used to pay for the items. But people sometimes pay cash as well as by credit card, sometimes use different credit cards, and sometimes different members of a family pay for items. Supermarket loyalty cards, however, provide a detailed record of which items are purchased by anyone in a family. It is possible to create a sophisticated profile of purchasing habits of a family: do they buy beer twice a week or 7-up five times a week? Do they buy crisps at the same time as beer? This can be used for designing better products, but can also be used to determine what other products someone may want, and may help them target consumer sales.

Suppose this information is sold to a third party, who can use it to target the family for advertising. Such information can be very valuable, especially when many organisations ask individuals to indicate their preferences, and people comply, in order to obtain the rewards offered by the organisations. Some supermarkets ask for dietary information about customers, others are interested in the number, age and gender of children. Some even ask for information about type of heating in the home, and number or type of cars owned (O'Dea 2001). Considering that each of the main supermarkets has about a million loyalty cards, that constitutes a large proportion of the Irish population. Large databases can be used to find correlations that help target consumers, linking particular products to particular consumer profiles (geographical, age, gender, leisure, etc), and then target the individuals who fit the relevant profiles.

What can organisations do with this information? What can other organisations do with the information, if it is given to them? Does the government have the right to know what is in a doctor's files on individuals? Do insurance companies have the right to know about individuals' traffic violations? Freedom of Information legislation enables citizens to have access to their own personal data, as well as public interest data, but it does not cover the activities of non-government organisations nor does it restrict what governments can do with information they have collected on citizens. There is an assumption that, while governments must keep information on individuals who break the law or who are a threat to the safety of others, the privacy of other citizens should be protected. However, the actual level of protection varies from government to government. Data Protection legislation determines how much information can be collected on individuals, what use can be made of that information, whether by a government department or private organisation, and whether than information can be passed on to third parties.

6.4.1 *Data protection legislation*

Data protection is recognised as a global issue, enshrined in international law. The 1948 Universal Declaration of Human Rights states: 'No one should be subjected to arbitrary interference with his privacy, family, home or correspondence, nor to attacks on his honour or reputation. Everyone has the right to the protection of the law against such interferences or attacks' (Article 12). Similarly, Article 8 of the European Convention for the Protection of Human Rights and Fundamental Freedoms 1950 states: '(1) Everyone has the right to respect for his private and family life, his home and his correspondence. (2) There shall be no interference by a public authority with the exercise of this right except as in accordance with the law and is necessary in a democratic society in the interests of national security, public safety or the economic well-being of the country, for the prevention of disorder or crime, for the protection of health or morals, or for the protection of the rights and freedoms of others.'

In countries around the world, the collection, use and dissemination of personal information by both governments and private organisations are subject to regulation, but laws about the protection of data vary from jurisdiction to jurisdiction. In many cases, there is a general law that regulates information in both public and private sectors, with an overseeing body to ensure compliance. This is the model adopted in the European Union and its member countries. In other countries, such as Australia and Canada, the private sector develops and enforces its own rules, which are overseen by a government agency. At the other end of the spectrum, the United States has individual laws for individual protection issues. These would include financial records, credit reports, video rentals, cable television, children's (under age 13) online activities, educational records, motor vehicle registrations and telemarketing. As new issues arise, new legislation has to be passed. In addition, the United States has generally favoured industry self-regulation, rather than regulation by a government agency. This has led to many criticisms of the adequacy of data protection in the private sector within the United States (for more detail on these issues see Electronic Privacy Information Center and Privacy International 2002).

Laws regulating some aspects of data protection have been brought into many jurisdictions since the early 1970s, for example in Sweden (1973), the United States (1974), Germany (1977) and France (1978). In 1981, the Council of Europe approved the Convention for the Protection of Individuals with regard to the Automatic Processing of Personal Data and the Organisation for Economic Cooperation and Development (OECD) approved Guidelines Governing the Protection of Privacy and Transborder Data Flows of Personal Data. In general, data protection legislation requires that personal information must be

- obtained fairly and lawfully
- used only for the original specified purpose
- adequate, relevant and not excessive to purpose
- accurate and up to date
- accessible to the subject
- kept secure
- destroyed after its purpose is completed

These principles are easier to enforce in countries such as those in the European Union, which have explicit agencies to monitor and enforce such rules across the wide spectrum of government departments and private organisations. It has been the case that, on the whole, data protection has been taken more seriously in Europe than the United States, with the rules as to what information can be collected and what use that information can be put to being more restrictive in the European Union, and with violations of those rules being more strictly punished in the European Union.

It is a feature of the information revolution that digital information does not respect national boundaries. Thus whatever rules for data protection exist within a country, private companies can circumvent those rules by storing the data in a different jurisdiction. Trans-border data flows are very difficult to regulate, and the best solution has been to restrict the transfer of information to third countries unless the information is protected in the destination country. Alternatively, organisations could be required to include a contract provision, when obtaining information from citizens, which binds the organisations to comply with data protection rules (e.g. right to notice, consent, access and legal remedies), regardless of where that data are stored. The most high profile case has been negotiations between the European Union and the United States regarding the rules with which United States companies must comply if they are operating in the European Union and obtaining data on EU citizens. The result has been a 'safe harbour' agreement, in which United States companies would voluntarily certify that they would comply with privacy principles worked out by the United States Department of Commerce and the Internal Market Directorate of the European Commission.

Until recently, privacy protection was increasing, with more countries throughout the world passing data protection legislation and with greater international harmony emerging to safeguard international (e.g. trans-border) data flows. In the aftermath of the events of 11 September 2001, it was believed that electronic communications (email, websites, and even digital graphics) could be used to co-ordinate terrorist acts. Even when there was no evidence that this had been the case in the 11 September events (Campbell 2001), there was a general move to relax personal privacy legislation to allow greater surveillance of electronic data communications. Almost every country has changed its laws, making it easier for government agencies to intercept communications and increase the range of data that can be accessed. Whereas it had previously been illegal in most jurisdictions to share information, there has now been a trend towards sharing data amongst government agencies. There have also been reversals on data retention policies. Previously, data could be retained for only a relatively short time, and little of the retained data contained personal information. There have now been suggestions that Internet Service Providers be required to archive all data (email, web access) for a year, in case security agencies want to apply for access to it. Some of the proposals extend the period to up to seven years (Lillington 2001a). The level of protection afforded personal data has diminished considerably since September 2001 (Mathieson 2002; Grossman 2002), although the benefit for global security of such a diminution has yet to be demonstrated.

6.4.2 *Irish data protection legislation*

In Ireland, data protection legislation requires that organisations storing information on individuals must register themselves with the Data Protection Commissioner. Individuals have the right to see the information about them, and have it altered if it is incorrect. The law limits those to whom organisations may give information, and what they can do with the information they have collected. By law, organisations must obtain and process the information fairly, keep it for only one or more specified and lawful purposes, use and disclose it only in ways compatible with the purposes for which it was originally obtained (unless permission from the individual is received). They must keep the information safe and secure, up-to-date and accurate. Equally importantly, the information can be retained no longer than is necessary for the specified purpose for which it was obtained.

These rules can have far-reaching implications. For instance, the Irish phone directory was made available in electronic form and it was possible to search for all individuals on a particular street. In 1999, the Data Protection Commissioner ruled that this was not the use for which the information was originally collected and was therefore in infringement of the individual's rights to privacy (Office of the Data Protection Commissioner 2000). More significantly, the Irish Data Protection Commissioner queried mobile phone companies about retention of call records for nearly seven years. The retention of these records turns a mobile phone into a 'tagging device', by providing a snapshot of an individual's daily life. This is because the 'locator records' of the mobile phone company continuously pinpoints a user's location to within a few dozen feet in urban areas, as the activated mobile phone moves from one transmission cell to another (whether a phone call is being made or not). Since nearly 70 per cent of Irish people own a mobile phone, this 'tags' most people. Mobile phone companies say that they keep records 'online' for only six months; after this, the records are archived, but are 'atomised' so that individual's name cannot be reconnected to a given set of records. However, mobile phone companies admit that company could reconnect records to individual names if need be, and would do so if required by law, which the Data Commissioner ruled to be illegal (Lillington 2001b).

Do organisations have the right to pass data on to third parties? Can a company sell its database of customers, with telephone numbers and addresses, to someone else? Can they sell their customer's buying patterns, based on records that the organisation holds? In the United States, the answer is yes. In the European Union, and especially in Ireland, the answer is no. Tesco is reported to be selling customer profiles from its Clubcard database to third parties (Harrison 2001), but Data Protection legislation means that Tesco sell only demographic patterns and consumer preferences, but cannot reveal names and addresses of its cardholders.

6.5 Regulatory issues

Freedom of Information and Data Protection laws are means by which governments regulate the collection and distribution of information. However, governments face many other regulatory challenges as well. This is especially the case for economic activities. As the rules of economic activity have changed, so must the laws and regulatory procedures change. This has sometimes led to conflicting imperatives. Firstly, individuals have a right to privacy. Individuals expect privacy in their communication with others, whether that communication is face-to-face, via letter, over the telephone, or electronic. Similarly, economic transactions may require private exchanges. If the transactions are electronic, the need for security becomes equally important. On the other hand, governments must also retain the right to override any security protocols, if the communication or transaction is for illegal purposes, such as drug purchases, money laundering, terrorist activities, or any other illegal activities. There is a conflict between the need for governments to monitor exchanges dealing with illegal activities while also permitting private exchanges on legal activities.

6.5.1 *Electronic commerce*

One can order goods online, pay for them online, and sometimes (depending on the good or service) obtain the good or service online. For electronic commerce, many requirements have to be satisfied: valuable electronic information has to be secure and safe from theft; when transmitted, it must be safe from disclosure to a third party; it must arrive without alteration; the sender must be authenticated (the sender is who he/she claims to be, and has the right to authorise a funds transfer or enter into a contract); only the intended recipient can see information, and there must be a legal basis for the testing of claims regarding such security, and a legal basis for enforcing any electronic contracts entered into. This is a tall order.

Traditional transactions were conducted in person, and the person's identity was either irrelevant (because payment was in cash) or else it was verified by signature or photograph. Signatures show that the person read the document, and is thus liable in court if the terms of the agreement are breached. Signatures are also used to verify identity when using a credit card, just as photo IDs are used to identify recipients of information. How can we verify identity in an electronic environment? As noted, there are limits to the dependability of PIN numbers – those four digit numbers, which, in combination with physical possession of the bankcard, are sufficient proof of identity for banking transactions. Many people write their PIN numbers down, and it is easy enough to steal both the card and PIN number and, by impersonating the account holder, withdraw cash. Banks recognise these security flaws by limiting cash withdrawals in order to minimise potential damage from fraud. For effective electronic commerce, greater security is required.

How does one ensure secure and verifiable exchanges on the Internet? The Internet was not designed as a secure closed system, and this is the precise attraction of the Internet as a medium for electronic commerce. There is casual access to it, and virtually any hardware or software can be used to access it. How, then, to ensure that information is not intercepted and copied, not altered, and authenticated in terms of who has sent it? The traditional answer is encryption: using a set of rules to alter data so that only the sender and receiver can decode it. This set of rules is called the 'key', and it protects the integrity of the data and verifies the identity of sender and receiver (since only sender and receiver know the encryption system and the password). Traditional cryptography is based on the sender and receiver of a message knowing and using the same key: the sender uses the key to encrypt the message, and sends the message (which is now composed of apparently meaningless letters and numbers) to the receiver. The receiver uses the same secret key to decrypt the message, and so recreates the original message out of the apparently meaningless string of letters and numbers. This method is known as secret-key or symmetric cryptography. A key composed of a few digits will lead to the same 'pattern' being repeated often enough for a computer eventually to break the code. However, if the key is long enough (128 length keys is becoming the standard), a message cannot be deciphered unless the recipient has the key. The main problem is getting the sender and receiver to agree on the secret key without anyone else finding out. If they are in separate locations, they must trust a courier, a phone system, or some third party with the secret key, all of which are potentially vulnerable. Furthermore, such a system does not work for electronic transactions, since the two parties do not necessarily have any contact beforehand, to share and protect the secret key.

There is one means by which messages can be encrypted and then decrypted without advance sharing of the key: public key encryption. The concept of public-key cryptography was introduced in 1976 by Whitfield Diffie and Martin Hellman (Diffie and Hellman 1976). In public-key encryption, everybody has two keys – a private key known only to the user, and a public key that is known to everyone and which is created from the private key. The sender encrypts a message using the recipient's 'public key'. Once encrypted using the recipient's public key, the message can only be deciphered using the private key that was used to generate the public key in the first place. Effectively, the recipient is the only person who can actually decode the message because he or she is the only one that has the necessary key. A reply is similarly encoded, using the public key of the original sender. Public-key encryption solves all the problems of authentication at once, because the process requires information that only the sender and receive have access to, without requiring them to share that information in advance. Since the message must be encrypted by the sender (using his/her private key) and decrypted by the recipient, it provides mutual authentication. It proves to the recipient that the message comes from the sender, it ensures for the sender that the message can only be read by the intended recipient, and it verifies that the original message has not been changed. This solution satisfies almost all the requirements of electronic commerce. As more computer applications (especially World Wide Web browsers such as Internet Explorer and Netscape) have the software for such public-key encryption built into the programme, it becomes easier for anyone to engage in such secure transactions.[2] Further developments are still necessary, such as an infrastructure to certify that an individual or organisation's public key is in fact from that individual or organisation, rather than an impostor. But, as this infrastructure develops and laws change so that digital signatures and public-key encryption have the same status as is currently accorded to traditional signatures and other means of personal identification, then secure electronic commerce (as well as electronic government) becomes possible. However, the very security of public-key encryption raises a different set of issues.

6.5.2 *Privacy*

Public key encryption enables electronic transmissions between individuals that not even governments can monitor, even if the messages can be intercepted (itself often very difficult). For most transactions, such privacy is not a threat. However, governments have always reserved the right to monitor and intercept communications on illegal activities, such as drug deals, child pornography, or terrorist activities. Telephones calls can be monitored, face-to-face conversations can be overheard, and postal mail can be intercepted. Governments vary in the procedures used to intrude into private communications, but they all reserve the right to do so. Public key encryption enables secure transactions that cannot be monitored; government agencies would need the sender and the recipient's private key to decipher the communication and it is unlikely they would have access to either. Encryption protocols would enable crime to flourish, as criminal activities can be carried out without governments being able to monitor communication content. For that reason, governments want the ability to intercept and decipher electronic exchanges, just as they have the right to intercept phone and mail communications. There is a conflict between those who want greater individual privacy (either to protect civil liberties or to enable secure economic transactions) and those who want less privacy so that governments can ensure public security.

This debate arose when the United States proposed that encryption be built into computer hardware. On 16 April 1993, the White House announced a new secure encryption standard, which became known as the 'Clipper chip'. While the encryption system was secure and fast (since it was hardware rather than software encryption), it was proposed that the US government be able to read any message encrypted on any Clipper chip, because it would retain a copy of the keys. The plan was to create a master database containing the serial number of each Clipper chip manufactured and that chip's master encryption key. The database was to be split between two 'escrow agents'. Law enforcement agencies could apply to each of the escrow agents for release of the key components, if they needed to decipher communications whose content was suspected of being illegal.

Would individuals trust governments to intercept only those messages that all would agree are illegal? Many civil liberties groups did not trust governments, and critics denounced the Clipper as Big Brother and refused to use it. They pointed out that criminals would not use equipment containing the chip anyway, and foreign customers would not want equipment that US spies could tap into, particularly since powerful encryption was available overseas. This would hurt telecommunications firms that sold their products abroad. In the end, the United States backed down over Clipper and exporting encryption, abandoning controls on the use of strong encryption in December 1999, as well as abandoning efforts to require 'escrow' for encryption keys.

However, governments, and especially law enforcement agencies, still argue for some escrow system for encryption. Secure electronic communication is a serious threat to security because, even having discovered illegal activity, governments would be unable to prevent such action, or prove, afterwards, that such action had been planned, because the communication could not be deciphered.

The demand for government access to private communication had lessened until the events of 11 September 2001. Concern for security, post-11 September, has increased demands for escrow for encryption keys and restricted access to strong encryption.[3] There have been proposals to make two significant and complementary changes in data protection and privacy rules. One of the current proposals relates to 'back door' access by governments to encrypted emails. This would enable governments to decode public key encrypted systems as part of anti-terrorism measures. This is linked with proposals for retention of all data transmissions, so that they could be examined at a future date, even though such retention would violate current Data Protection rules.[4] The two, in combination, would enable tracking of many activities, movements and communications of individuals. However, it might also diminish the attractiveness of countries with such rules as locations for e-commerce, since many businesses would not trust that their communications would remain secure. The issue continues to be debated and it arises also in discussions on censorship.

Public key encryption not only ensures the integrity of data, but it also provides a digital signature which is necessary as a means of validating individual identity so that a contract (sales transaction, for instance) entered into electronically entails a subsequent legal obligation. However, a technical solution will have impact only if backed up by law. Yet it is difficult to frame laws in this area, as it means incorporating some very technical procedures into law. There is a common-sense understanding that only a person can provide his/her own signature and procedures exist to detect forgeries, so it is relatively easy to write laws that can be enforced in courts regarding personal signatures. It is much more difficult to demonstrate the validity of the set of technical specifications that create a 'digital signature'. The courts are expected to make decisions on technical rather than common-sense criteria of identity, without themselves being technical experts and sometimes having to listen to conflicting views from different experts. This is a difficult issue to decide.[5] It is also difficult to convince the public that such a system actually works. Yet such legal protection for encryption and digital signatures are crucial for electronic commerce, and governments will have to bring in appropriate laws for these activities.[6]

6.5.3 *Copyright and intellectual property*

Electronic transactions require privacy, but electronic content must also be protected in the new information economy. There must be laws to protect information as a commodity, just as there are laws to protect any other 'property'. It is relatively easy to protect physical objects – one simply prevents them from being removed, or prevents other people from exercising control over them. But information is not an artefact, and it can be removed or accessed much more easily. As with any commodity, the first step is to protect the item, which means secure storage of information. This involves restricted physical access to computers, but also firewalls and password protection to restrict electronic access. Since information has to travel from producer to consumer to be valuable, there also has to be secure transmission, usually through some encryption system.

Theft of information, which would mean unauthorised access and copying of information, is inevitable. Copyright law regulates who has the right to copy and use information. For non-digital information, the cost of making and distributing copies of information both restricted the amount of copying that went on and also made it relatively easy to control copies. One could monitor the expensive technology required to mass produce books, magazines, records and film. As copying technologies became cheaper, control became more difficult. Xerox machines and video recorders enabled individuals to make copies without concern for copyright laws or payments. However, the loss in the quality of the reproduction and the cost in making copies limited the amount of copying. Digital information changed all this; it is easy for individuals to make cheap and accurate copies of digital information with very little capital investment. The result has been illegal CDs, pirated software and music exchanged via Napster (or some other free information exchange). As soon as a record is released, there are free versions which individuals can download and then write onto a CD; as soon as a film is released, there are digital versions available on the Internet which can then be turned into home DVDs. How can 'ownership' be protected, so that owners can extract payment for intellectual property? One solution is to prevent copying of digital information, and there is an enormous amount of research devoted to making such reproduction difficult, if not impossible. Newer software programmes are designed to play only 'legal' audio or video data; some new CDs will not play on personal computers (so as to avoid copies being made).

Despite such efforts, there remains a thriving illegal software industry. There must be laws to enforce intellectual property rights and ensure that owners of copyright receive payment when others use their material. Increasingly, one does not pay to 'own' information, one pays for the right to access or use the information and ownership remains with the provider. The 1998 Digital Millennium Copyright Act (DMCA) was passed in the US and the European Union Copyright Directive is being passed into the law of EU states, both of which regulate intellectual property and digital information. A century ago, copyright law offered only 14 years of protection; it now extends 70 years after the creator's death. Thus, this new intellectual 'property' can actually be inherited, bought and sold. This may create unexpected restrictions, since if a company or individual who has copyright does not wish to release it (it may be profitable enough), no one else has access to it either. Information may thus become inaccessible.

It is the owners and distributors of information (often the large multinationals) who are promoting laws that impose the same control over digital information that was previously exercised over non-digital information (such as vinyl records or books). There are many other individuals and groups who wish to exempt digital information from copyright restrictions. Public domain information is available to anyone free, and computer programmes as well as digital content are sometimes made available in the public domain. Linux, for example, is an operating system for personal computers that is freely available, an alternative to Microsoft Windows, and Open Office is a freely available suite of computer applications that provides functionality similar to Microsoft Office.

6.5.4 *Consumer rights and taxation*

Copyright and intellectual property have been redefined so that information can be an economic commodity. When this commodity is purchased, what rights do consumers have? How does one apply consumer rights when electronic purchases span different jurisdictions? What does the consumer do if the product does not match advertising, but the seller resides in a different jurisdiction? Or if the seller cannot even be identified, because the web page provides electronic addresses but not physical addresses?

Within the European Union, the regulations of the selling country apply in regulating purchases. An Irish resident who purchases from a French seller will find that French consumer protection laws apply. Increasingly, there is a harmonisation of rules; for instance, all EU countries require accurate descriptions of goods and for refunds to be made within a specified period of time. However, this only works if countries have agreed rules that apply across different jurisdictions and if it is relatively easy to bring sanctions to bear when the rules are broken. In many cases, trying to exercise one's rights in a different country is such a nightmare that few cases are brought, even within the European Union. Many of those who sell goods do so in countries that do not agree to such international conventions.

One thing that governments do agree on, however, is the need to collect taxes. In most jurisdictions, economic transactions are taxed, whether by a sales tax or valued-added tax (VAT). In the United States, sales tax is paid in the jurisdiction of the consumer, so, if you buy something in the state of California, you pay Californian sales tax, if the item is ordered and sent within

California. In the EU, VAT is paid at point of purchase, even if the consumer then takes the item to another EU state. In both cases, there is a physical item that can be tracked and a tax levied. But what about electronic transactions, when the seller and buyer are often in different jurisdictions? At the moment, even talk of such taxation is academic, as the transaction is often impossible to monitor. If the item is delivered electronically (e.g. digital file, whether that be music, a computer programme or a movie), it is very difficult to track the transaction, much less determine whether it should be taxed at point of production or consumption, or neither. This does not mean that governments are not even now trying to develop systems for taxation, but it does mean that any solution depends on international agreements that are still some years away.

6.6 Censorship

Discussion has focused on the regulation of information as an economic commodity, but information has other impacts in addition to economic ones. In popular discussion, one of the terms most often associated with the Internet is pornography, especially when police arrests of individual offenders attract mass media publicity. Pornography is seen as a phenomenon of modern technology, although there is little evidence to support this (Rosenberg 1993; Burton 1995). Certainly, however, one feature of the digital revolution is the ease with which information flows from place to place, and the ease with which individuals (especially children) can privately access inappropriate information. Should (and can) states monitor and restrict the movement of information? Governments have traditionally exercised control over information within their jurisdictions, which, to be effective, required the ability to control information entering from other jurisdictions. If it was illegal to publish the information in the jurisdiction (e.g. information on abortion or contraception), then it was also illegal to import the information into the jurisdiction. In that past, such control was easy. Within the jurisdiction, traditional publishing was controlled through taxation, regulation and censorship. This was effective because of the large capital costs required to publish newspapers, magazines and books as well as to broadcast television programmes, and the threat to confiscate such investments or put the individuals involved in such investments in jail. The distribution system for printed materials and electronic media was also easy to regulate. It was equally easy to control information coming in from outside the jurisdictions: it was encoded on physical media (books, magazines, video tapes or tape cassettes) and could be stopped at a physical border. Even electronic information could be controlled, as information from outside the jurisdiction could not easily reach consumers, unless there were local relay stations that could, themselves, be controlled.[7]

Digital information flows are not so easy to regulate or even monitor. Gone are the days of receiving illegal publications in brown wrapping or buying them under the counter at a bookseller or newsagent in order to minimise detection. Using new technologies (including, but not restricted to, the Internet), individuals can access illegal information in a much more private and unmonitored environment. Electronic broadcasts via satellite require only a small dish for reception of the signal, and the producer of the broadcast is outside the state's jurisdiction. Digital publications can be based outside the jurisdiction, and it is virtually impossible to examine electronic transactions, whether they travel via dedicated data lines or telephone lines. If the information is sent using the Internet, the data are divided into packets that travel separately and are reassembled at the destination point. Unless all data traffic into a country is directed through the same gateway, it is virtually impossible to monitor any transaction. In any event, free flow of information, capital and labour are necessary for participation in the global economy. If a state is too restrictive in its control over the flow of information, then multinationals will locate elsewhere. Most countries consider a loose control of information to be a price that has to be paid for economic participation.

While the definition of 'illegal' content varies from country to country, there is some information content that almost all jurisdictions consider illegal. This would include child pornography, racist propaganda, information on creating weapons and terrorist-related information. What are the mechanisms by which states can control such illegal information entering their jurisdiction?[8] There are four options: control at point of information provider or publisher; control at the national level of links to the outside world; control at level of Internet Service Provider or organisation (if accessed at work), or consumer control (see Government of Ireland 1998; European Commission Legal Advisory Board 1996). The digital information chain from producer to consumer is a complicated one and often spans different jurisdictions. A publisher can live in one jurisdiction, upload the content to a server that is located in a second jurisdiction, while the consumer lives in a third jurisdiction and accesses the material via an Internet Service Provider in a fourth jurisdiction, and the data can pass through a fifth jurisdiction. If the different jurisdictions have different rules, in which jurisdiction do the rules apply?

6.6.1 *Source of publication*

When information is published in one jurisdiction and read in another, do the laws where the publisher is located or where the reader resides obtain? In traditional publishing, both jurisdictions apply, but electronic publishing is more complicated. Child pornography is illegal in the United States and European Union countries. However, evidence from the United States is not automatically accepted in other jurisdictions. When a site selling child pornography in the United States is raided, the records of customers from other jurisdictions can then be used as the basis for investigation but independent evidence must be found in the customer's jurisdiction. This has been the basis for many of the recent prosecutions for child pornography, where police from different jurisdictions co-operate in order to trace the distribution and use of illegal material.

However, some activities are illegal in one jurisdiction but legal in another. It is illegal in France to publish Nazi material or sell memorabilia, and a US web server was brought to court for hosting a site on which Nazi memorabilia were sold (Kelleher 2001). Not only is such an activity legal in the United States, but it is actually illegal to prohibit such publication since the First Amendment in the United States constitution forbids such restrictions on free speech. The web publisher argued that, even if they were willing to comply voluntarily with the French ban, it was impossible to apply French laws to American information providers in such a way as to affect only French Internet users. In the end, the French court ruled that, in so far as was possible, the US provider should prevent French users from accessing the site. There was little, in practice, that the French courts could do to enforce such a rule, other than act against any French subsidiaries of the organisation.

In a similar case, CompuServe in Germany was brought to court in the mid-1990s for allowing access to Usenet discussion lists that included pornographic and Nazi material. The court held that the head of German CompuServe was liable for allowing open access to Usenet; however the guilty verdict against the manager was later overturned. The local representative of CompuServe was held not to be responsible for the larger organisation, and it was agreed, even by the Bavarian prosecutors who raided CompuServe offices in December 1995, that there was no technology available at the time that would have enabled CompuServe to block Internet content.[9]

One answer would be to ensure that all provider host countries operated the same rules within their respective jurisdictions, so that what was illegal in one jurisdiction would be illegal in the other. International conventions in which sovereign states agree common legal definitions of crimes provide a partial solution to the problem of regulating content. Thus there has been formal adoption of a 'cybercrime convention' by Committee of Ministers of the Council of Europe in November 2001 (Smyth 2001). It criminalises some computer-related activity and seeks to harmonise laws across 43 Council of Europe member-states, the US, Canada, South Africa and Japan. The convention covers many regulatory issues, such as infringement of copyright, computer-related fraud and child pornography. Such international conventions will proliferate over the next few years, enabling governments to control publication at source. However, there will always be some jurisdictions with lax regulations and publishers can relocate to those jurisdictions. This applies to sites that publish racist, pornographic or any other illegal material. Relocating publication to 'safe haven' jurisdictions may prove an inconvenience,[10] but it will always remain an option and so controlling publication at source will never stop those who are determined to publish.

6.6.2 *Borders*

If it is not possible to prevent publication at source, is it possible to exercise control at the interface between publisher and reader? In the case of the Nazi memorabilia, it was suggested that the US provider block access by French IP addresses (at least to those segments of the site which display material that is illegal in France). Although there are numerous technical reasons why such an attempt is likely to be only partially successful,[11] it does suggest a strategy by which access to sites that have not agreed to international conventions could be banned. Thus, if a country has not signed the relevant convention, then the local Internet Service Provider (which is the way most people access the Internet) could prevent access to any sites hosted in that country. Many would consider this an excessive restriction on personal liberty. In any event, there would be problems in exercising such controls effectively, since it would require blocking large numbers of Internet addresses indiscriminately.

Some governments control Internet content by restricting the number of ISPs in a country and then monitoring the activities of those that do exist.[12] This can also reduce the number of people who can access the Internet, since the capacity of the ISPs is limited. Another means of state censorship is to impose charges that make it too expensive for most citizens to access the Internet. Some countries are even less subtle, and simply reduce the level of connectivity between users and outside world to the point at which Internet access is impossible. Such measures impede economic growth, but this may be an acceptable cost for countries whose leadership feels threatened by free flows of information (see also Taubman 2002, http://www.firstmonday.org/issues/issue7_9/; Van Koert 2002, http://www.firstmonday.org/issues/issue7_4/ ; Kalathil and Boas 2001, http://firstmonday.org/issues/issue6_8/kalathil/).

Can Internet Service Providers be responsible for illegal information which their customers might access? There is disagreement whether ISPs are publishers or common carriers. Newsagents, publishers and broadcasters are responsible for the contents of what they sell, publish, or transmit, because they are deemed to be able to monitor the contents of what they sell or broadcast. Telephone companies and postal companies are deemed to be common carriers; they are not aware of the contents of the material that they carry and are not expected to exercise editorial control over that content. Into which category should ISPs be placed? Should, and can, those who provide Internet access be expected to monitor and control data transmissions?

Clearly, ISPs can be held accountable for material that they themselves host on their own servers. But how can they monitor what material is accessed by their customers on sites elsewhere? This could only be done by monitoring reassembled packets and restricting which sites or content individuals can access. This would be very unpopular with users, would severely undermine the benefits of Internet access, and would be difficult and expensive to enforce. It is virtually impossible to discover the addresses of illegal sites and, once discovered, such sites can simply change address. The most that one could do is keep a log of IP addresses which users access; this could be used in future prosecutions if charges are ever brought by police regarding individuals who were accessing illegal material. This depends on monitoring individuals and their use, rather than a blanket ban exercised at the interface between publisher and consumer (see also Burton 1995; Rosenberg 1993). In the context of increased security since 11 September, such measures are being proposed, but many civil liberties groups oppose such intrusion into private life.[13]

6.6.3 *Consumer control*

The most effective control is exercised at the point of reception, that is by making it illegal for individuals to store certain information (including images) on a computer or to access information (e.g. reading but not storing information). This takes into account the reality that it is difficult to prevent individuals from accessing illegal material on the Internet, given the international nature of Internet publishing. There are two types of control at this level: one is to restrict access by some but not by others (e.g. preventing children from accessing pornographic sites which adults are permitted to access) and the other is to prohibit all access to sites that host illegal material. In the first case, control is imposed voluntarily, usually by parents or organisations. The most common control is installing monitoring software that either monitors or blocks content. Such software uses three different strategies: exclude everything by default, unless enabled (which prevents access to new sites); include everything, unless disabled (which requires knowledge of all sites to be excluded, including new sites that develop); and filtering software that checks for content on a case-by-case basis. The problems with such software are well known; they either block too much, or too little, or they are too easy to get around. For the personal user, such software works only if the user chooses to implement the software, but many families would not know how to do so. It does, however, prevent illegal use by individuals in organisations, as organisations possess the expertise and resources to implement software and maintain a list of blocked sites.

A different sort of solution, which is less dependent on users, is to implement content certificates for sites and to implement software in all World Wide Web browsers that would require such certificates. The user simply notes that a certain level of censorship is to be implemented, and leaves the implementation of this to the computer operating system. This solution also has its own problems, since it assumes consistent rules for classifying content, that sites will implement such rules, and that users will use the software for such monitoring. There is no immediate prospect of software that can classify images as 'pornographic', and checking word content sometimes excludes harmless material. However, it offers some level of control in circumstances where parents have less computer expertise than their children.

Controlling the content on the user's computer assumes that individuals want to implement such controls. This may be the case for parents who wish to protect their children and organisations which wish to protect themselves against criminal charges. But much activity is undertaken by individuals who want to access illegal sites (whether child pornography, racism, or terrorism). Usually, individuals are detected by tracing who has been accessing illegal sites, tracking them back to the Internet Service Provider, determining which individual was accessing the sites, and then seizing the actual physical computer to find evidence of illegal activity. It requires both publishers and consumers of such illegal content to be identified, even though such exchanges take place in multiple jurisdictions. It is clear that control of content, if possible at all, will involve multiple layers and approaches, and there is unlikely to be a permanent and clear-cut solution.

Most illegal activity is probably undetected because it is relatively easy to publish illegal material without being detected and it is easy for users to access material without detection. The effective solution would be to impose monitoring software on all computers as a legal requirement. This is not a solution that would find favour with many citizens, as it would require a fundamental intrusion into personal privacy. Furthermore, it avoids the question of who decides what constitutes illegal information. Cultural definitions of 'pornography' vary from one society to another, and attempts to create legal definitions that are acceptable across jurisdictions have not been successful. The history of book and film censorship shows how definitions vary from one society to another, change over time, and are even the subject of disagreement within societies (Couvares 1996; Williams and Great Britain Committee on Obscenity 1979; Barker, Arthurs et al. 2001; De Grazia and Newman 1982; Phelps 1975). Furthermore, even if many jurisdictions consider child pornography illegal, some also consider information critical of a current government to be illegal. Definitions of 'legal' content are political as much as social, and permitted dissent in one jurisdiction may be illegal subversion in another. The regulation of public information (whether printed books and newspapers, electronic broadcasts, or digital Internet-based) has always been affected by the changing political concerns of governments and cultural values of society; new technologies have raised public awareness but not fundamentally changed the issues.

Chapter 7

Political participation

7.1 Introduction

In earlier chapters, the role of states in promoting and regulating economic activities has been discussed, as have government policies regarding the regulation of information. Who decides what states should be doing in these areas? Who decides what information should be protected under data protection legislation, or what the rules of electronic commerce should be? These issues of policy have to be decided before governments can administer those policies, and all citizens play a role in determining public policy. The potential political consequences of new technologies are significant, possibly changing the way in which citizens participate in policy formation.

For citizens in democracies, political participation in policy formation is restricted to voting at elections, choosing amongst the competing policies of different political parties. The parties with the most votes then have the mandate to bring those policies into force. People may also vote at a referendum if the matter is something that elected politicians are not permitted to do themselves, and which may require a change to the constitution of a country. There are also a range of informal mechanisms by which people have input into policy, from writing to their politicians to participating in interest groups that lobby governments and politicians on specific issues. All of these are means by which people can influence decisions about the allocation of the scarce resources (often tax revenue) controlled by the state.

The democratic process is changing as a result of new technologies, and there is potential for greater change in the future. Electronic voting, virtual public meetings, politicians participating in Internet question and answer sessions are examples of current political changes. The structure of traditional political parties is changing, and new political parties and interest and community groups are emerging, to enable citizens to participate in policy formation in new ways. There is a tendency towards a displacement or marginalisation of formal political parties; interest groups, social partners and community groups tend to bypass the political parties and articulate policy concerns directly to administrators and political decision makers. New technologies have been facilitating such trends by providing inexpensive means of communication and information transfer, so that small groups find it easier both to organise internally and communicate externally.

Many countries have experienced increased political apathy, with diminished interest in voting and political participation generally. Information technologies provide mechanisms by which such a decline in participation might be arrested and enhanced political participation encouraged. Many governments have issued policy documents proposing such improved participation, either through traditional or newly developed structures (Melody 1996). These range from the 'National Information Infrastructure' pronouncements in the United States (Gore 1991; Information Infrastructure Task Force 1993) to the European Union's High Level Expert Group on the Social and Societal Aspects of the Information Society (Commission of the European Community 1996) noting that 'ICTs create new opportunities for greater public participation in and awareness of the political process'.[1] The report by Ireland's Information Society Steering Committee (Information Society Steering Committee 1996) suggests similar policy directions, projecting that 'Government will become more accessible and responsive to its citizens' needs'. These policy documents paint a picture of a future in which citizens are informed about current issues and participate in policy decisions as interested and concerned parties. By and large, these documents are aspirational: what is the evidence for such changes?

7.2 Representative democracy

To understand why people expect new technologies to revolutionise and transform political structures, one must understand how contemporary democracies developed. For many industrial democracies, the model of government is representative democracy, and people imagine no other way for a democracy to function. But representative democracy developed as a compromise between the desire for citizens to contribute to policy, and the constraints imposed by the limitations of distance and size. Before the emergence of states, political decisions were consensual, based on discussions in which everyone participated (Roberts 1979; Lewellen 1992).[2] Even after the emergence of states, it was still possible for all citizens to meet together. Policy issues could be discussed and decisions taken on issues affecting the entire political community. This model of a forum for public discussion is taken from the *agora* of the early states of Greece where all citizens (but neither slaves nor women) could meet (*Encyclopaedia Britannica*, 2003), and this can still be the case in rural small towns or villages. This is direct democracy, and for many it is the ideal model for policy formation. However, it requires an informed citizenry who can meet, on a face-to-face basis, over a sufficient period of time to come to agreement (whether by consensus or a vote). As the size of a political community grows, direct input becomes impractical. It is impractical for all citizens to meet together for long enough to make policy decisions. Imagine the size of a town hall that would be necessary for all the citizens of even a small city to meet together, and imagine how long it would take for citizens to inform themselves of the issue involved, discuss it, and then come to a decision. It is simply too expensive to inform all citizens of the issues involved, too costly for them to travel far enough to meet together, and too difficult for them to gather together in one place where they can debate and discuss and vote on the issue.

Once a political system grows too large, representative democracy becomes a necessary substitute for direct democracy. Citizens elect someone to represent their views, and delegate to these elected representatives the power to make policy decisions on their behalf. While elected representatives sometimes consult with their constituents, it would be time-consuming to obtain their views on every issue, so politicians make their own decisions. Their policy positions would have been endorsed by voters at the previous election, so they could then make the decisions that they believed their constituents would desire, or would be in their best interests. At subsequent elections, the electorate could change policy by electing someone with a different ideological or policy stance. While it is possible for all citizens of a nation to vote on the same set of politicians, it is more common that a country is divided into a number of regional constituencies, and for people to decide amongst the competing candidates within that constituency. The assumption is, then, that the politician will articulate the concerns of that locality or constituency.

Although representatives are elected in local constituencies, they are also members of political parties which operate at a national level and which articulate broad policy orientations. The process of elections, as well as parliamentary discussions, are the public fora in which people and groups with differing views debate and, eventually, accommodate each other to come to a consensus view on collective policy. In addition to elected representatives articulating the concerns of a locality, interest groups articulate the concerns of particular occupational or other groups, such as doctors and environmental activists. These groups actively contact politicians or civil servants to exert pressure on those making policy, and thus are often referred to as 'pressure groups' or 'interest groups'. Politicians or bureaucrats may voluntarily consult these groups while considering policy decisions to ensure as broad a consensus as possible.

More recently, the policy process in many liberal democracies has also come to involve 'social partners'. This has been an attempt to create consensus politics among the various interest groups, including trade unions, employers, and some generalised vision of the 'public'.[3] Social partnership is founded to some extent on the pragmatic realisation that voluntary compliance by all members of society is necessary: compliance cannot be imposed or coerced, as there are too many informal or unofficial means by which resistance can be expressed. If all members of society, or their representatives, are involved in policy formulation, they can be held accountable for any non-compliance in the execution of policy.

All of these interests and groups form a 'public sphere', as Habermas (1989) and others have argued, which has been crucial for rational-critical debate in the formulation of public policy, as new mass media have permitted a larger number of people to become involved in policy debate. Newspapers and pamphlets were vital in encouraging discussion, changing people's views on various issues, and increasing public support for political positions. Pamphleting has often been a crucial means of spreading alternative political views throughout the world. More recently, newspapers, radio and television have also provided important vehicles for communicating policy issues and encouraging debate. These are obviously important because of the reduced cost of producing the information, in terms of the number of people that can be communicated with, as well as reducing the time delay in reaching them.

These technologies permitted more information to be communicated to citizens from established political groups, but such mass media outlets tend to exclude alternative political views, and alternative policies have tended to be marginalised in policy debates. Furthermore, while information could be distributed to the general population, only a small portion of that population was able to participate in the formulation of policy. The structure has not necessarily been the best possible, but it has developed to accommodate the practical limitations of communication and participation in ever larger political systems. It is impossible to underestimate the extent to which the information technology revolution has removed the previous restrictions on information flow as well as communication and transportation. The fundamental reason for representative democracy as a substitute for direct democracy has now been removed, leading people to rethink the fundamentals of how people participate in politics and policy.

7.3 New political structures

Representative democracy has been based on the unavoidable limitations of previously existing communications technologies, but new communications technologies reduce communications costs and increase speed of communication. Direct, rather than representative democracy, is once again possible, despite the large numbers of citizens and the large geographical expanse involved. It is feasible for citizens to inform themselves on a wide range of issues, and participate in discussion and debate, make their opinions known in the decision-making process, and even vote directly on issues. Even if the structure of representative democracy is maintained, new mechanisms of collective participation are possible. Local groups and interest groups can meet and discuss issues, convey their opinions to policy makers (politicians and civil servants) and monitor the application or administration of policy as it affects them. Just as new technologies have led to 'disintermediation' in many organisations (with intermediaries between levels of an organisation disappearing), so also could politicians, political parties, and interest groups diminish as links between citizen and state.

7.3.1 *Traditional political parties*

Traditional political parties have been affected by new information and communications technologies in the same ways as other organisations. New technologies are being used to reduce communication and co-ordination costs with external audiences and internal activists. They have been used to reduce organisation costs, obtain better information on public concerns, and target specific electoral interests. Political parties use websites and electronic newsletters to mobilise existing supporters and recruit new ones. New technologies are used to conduct research on policy issues more effectively, and, by using detailed demographic information on voters, to target specific kinds of voters with specially designed mailings or advertisements that appeal to class, regional or other special characteristics. It can lead, for example, to giving speeches tailored to the concerns of voters in a particular area. Many politicians use computer technology to keep track of constituents with whom they have contact; such records can be used to seek political support at elections.

Elections have also been altered by new technologies. They have increased the efficiency and reduced the cost of opinion polls, so that political parties can gather, input and process survey data at great speed and little cost. During elections, political parties commission private surveys to gauge public opinion and alter policy in response to public opinion. This has made political parties more responsive to public input than they previously were. During elections, political parties also use new technology to co-ordinate activities, especially policy statements. As issues arise, research can be done quickly and then disseminated to all the candidates. This means that all the candidates can respond in the same way to the same issue, even if that issue arose only in the last 24 hours.

New mass media technologies have been largely viewed as a new means by which the electorate can be targeted; it is an extension of other mass media communication, which is a one-way communication from the party to the electorate. However, new technology can also be used by citizens to contact politicians, which facilitates debate and discussion. Political parties as well as politicians in many countries now have email addresses as well as response forms on web pages, so that voters can make suggestions or respond to statements from political parties and politicians. These communication channels are a supplement to the traditional ones of personal visits and phone calls, although there is some evidence in the United States of new communications technologies replacing traditional means of mobilising supporters. Howard Dean, the early front-runner in the contest for the Democratic nomination in the 2004 US Presidential election, depended on new technology in his campaign for the nomination. He used a website to raise two thirds of his initial funding of $7 m., and supporters used geographic-based software to enable people to find other supporters in their local area, so that they could meet face-to-face and organise themselves locally (Butcher 2003; O'Brien 2003). These tactics brought in significant cash contributions[4] and mobilised a large number of potential supporters who had not previously been politically active. While this support was confined to a small segment of the electorate and Dean did not attract sufficient popular support to obtain the nomination, the campaign demonstrated the potential of new technologies to mobilise voters inexpensively and this will be emulated by other politicians.

7.3.2 *Special interest groups*

Some social and political issues are of passionate concern to a small number of individuals, but of only marginal interest to other most people. These individuals may organise into single-issue organisations (as opposed to traditional political parties, which have policies along a broad range of issues). Some examples of single-issue organisations would be ecology, nuclear disarmament, and refugee rights groups.[5] In the past, individuals interested in such issues could not undertake collective action because they were too dispersed geographically to co-ordinate their activities or exert pressure on political parties or policy makers. The communication costs, both internal and external, to support such activities would have been too expensive, given the small numbers and large distances involved. The best they could hope to do was to bring such issues to the attention of the main political parties and try to alter the policies of such parties.

New technologies have brought about an increased number of interest groups and also enhanced their public impact (see Melucci 1996 for a discussion of social movements). Geographical dispersion is no longer a barrier to effective political communication and co-ordination. Where activists in traditional parties meet in local areas to co-ordinate activities and share information, those in fringe groups can communicate electronically and act in concert. Individuals who would previously have been isolated and marginalised can have an impact out of proportion to their numbers. They can communicate with the general public by creating a website, making videos for distribution to television news, providing material for cable or satellite broadcast, or electronic publishing. They can stage 'events' which will then be covered by the mainstream media, thus reaching a large audience. The ecology and nuclear disarmament movements are both prime examples of dispersed individuals engaging in collective action. Ecological protests, when reported by news programmes, raise public awareness and put issues on the political agenda that might otherwise have been ignored. This, in turn, leads to mainstream political parties responding to public concerns, as illustrated by the protests in Ireland in 1999 over genetically modified organisms (O'Sullivan 1999). These activities can, and have, set political agendas for politicians, political parties and governments as a whole.[6]

7.3.3 *Community politics*

Much of the discussion of political change as a result of new technologies has focused on the use of new technologies for local empowerment, involving both greater participation in local decisions by all as well as participation in national decisions by vocal local groups (e.g. Carter 1997). New technologies could create electronic town halls, so that all citizens could participate in collective discussion and decision making. Individuals could find out about current policy issues, and contribute their own views as to what the city or local government organisation should do. The Santa Monica information system (PEN), in 1989, is often cited as an example of community networking, but it has encouraged little public participation and has been used to improve administration rather than public policy deliberations (Docter and Dutton 1998). There have been a number of other developments in community networking (for examples, see Tsagarousianou, Tambini et al. 1998). The FreeNet movement (sometimes also known as Community Networking) has been active in the United States and Canada, and is a prime example of local activists using new technology to increase participation in local community policy making. These movements receive varying amounts of state support, very little in the United States, except for some tax relief, while the Canadian government supports a community access programme (http://cap.unb.ca/). In Ireland, while there is evidence of communities using new technology, this is still mostly for encouraging extra tourist revenue.

The high infrastructure cost (phone charges, capital costs of computers and modems), as well as a general lack of awareness of new technologies, remain impediments to community networks, although efforts are being made to encourage community networks throughout the world (Huysman, Wenger et al. 2003). The main impact of new technologies in community politics is the ease with which individuals can organise when specific issues arouse great local interest. These tend to be short-term issues, and, owing to the informatisation of government administration, such groups can communicate electronically with both politicians and administrators. Individuals can organise ad-hoc meetings and exert pressure on local politicians and officials, which enhance local community solidarity and common identity. Even if such groups do not last long enough to enable community networks to develop, they still make greater political participation possible.

7.3.4 *Direct democracy*

At the moment, citizens can make direct policy decisions only on issues that are beyond the remit of elected politicians, for example with regard to changes to a national constitution which require the direct approval of citizens through a referendum. New technology could enable citizens to have a similar direct policy input on every issue; in a participatory democracy, all citizens would participate in decision making. Governments are currently experimenting with the more modest aim of introducing electronic voting into traditional elections. The major motive is to encourage greater electoral participation by making voting easier, although added benefits would be to reduce administrative costs and improve voting accuracy (Dutton 1999:173–93).[7]

Electronic voting can take two forms. At its simplest level, it involves people going to physical voting booths, and satisfying the normal requirements about eligibility to vote. They then vote using an electronic machine rather than a physical vote. The benefits of such voting are faster counting of votes, with less scope for human error. In the case of complicated voting systems, such as that which exists in Ireland, it would mean election results in hours rather than days with no need for recounts. The Irish electoral system uses multi-seat constituencies with proportional representation for elections. This system makes it possible for political parties with twenty per cent of the vote (or even less, depending on ranking voting preferences) to elect candidates to parliament, but it also means that elections to the national legislature can be decided on as few as 20 or 30 votes; in the May 2002 election, in one constituency only one vote separated winner from loser (Kennedy 2002). Electronic voting will mean faster and more accurate tabulation of votes.

Electronic voting can result in voting being extended to new areas. New technologies were used to great effect in South Africa in 1999 (Cross 1999). Polling stations were linked electronically (including satellite links) so that results could be collected, despite the poor communications infrastructure (especially the roads) that would otherwise have delayed counting of voting by up to a week. In the 1999 elections, 14,500 polling stations sent results to Pretoria in triplicate: phone, fax, and a high-speed data network. The same results had to arrive by two media, independently, in order to be verified. By virtue of speeding up the vote count by a week, it improved the legitimacy of the poll, and, now that the network is in place, will allow greater electoral participation at less cost than before.

A more elaborate system of electronic voting would involve citizens voting electronically, as well as having their vote tabulated electronically. This would reduce the costs of holding referendums, since people would use the existing communications infrastructure to vote, thus making it affordable for governments to hold many more referendums. However, this would involve many complexities. A system of validating identity electronically is required, providing inexpensive access to electronic voting systems and to ensure that votes are both confidential and secure. Such electronic voting is still in the experimental stage. In recent local elections in the UK, both Sheffield and Swindon used electronic systems that permitted people to vote from home, from public Internet sites, over the phone, and even via SMS or text messaging (Mathieson 2003). One of the aspirations of the local councils was that if it were easier for people to vote then more people would do so. The experiment was a modest success, as voter turnout remained the same even though turnout throughout the rest of the country declined by five per cent. It remains to be seen if electronic voting would reverse the decline in electoral participation.

If these experiments are successful, it might mean that electronic voting could not only be used in traditional elections, but also as a means of obtaining policy input on a regular basis. Instead of politicians making the policy decisions which they think their constituents would wish, the constituents could actually vote on the policy issue. Not a few political scientists (as well as many politicians) would argue that such direct decision making would be contrary to good governance. Busy citizens could not be expected to be informed on every issue, and if those who lacked sufficient time or expertise were expected to make decisions about complex laws, then the result would be bad laws. That is why full-time politicians with their own staff of specialists are necessary, along with the whole panoply of consultative interest groups. Direct decision making might also polarise and factionalise the electorate; elections tend to be conflict based whereas the legislative process tends to focus on consensus building. In any event, there is little evidence that voters are interested in such direct democracy. There are only a limited number of issues on which voters seem to feel strongly enough to want a policy input. For the rest, they are content to let others (political parties, interest groups) make the decisions (see Birrer 1999).

New technologies are, however, increasing public input into legislation in other ways. With new computer technology, surveys of public opinion are easier, cheaper and faster, with a consequent increase in their number. In addition to house-to-house surveys, there are now mail surveys, telephone surveys, and web-based surveys. Sometimes these surveys are carried out by political parties, sometimes by politicians concerned to discover local opinion on an issue, and sometimes by mass media organisations, such as newspapers or television stations. The surveys permit politicians to gauge public opinion more frequently, and on a wider range of issues than formerly. As politicians feel increasingly vulnerable to public opinion, frequent surveys are leading to a more responsive political system, even without the need for direct input by citizens on specific laws. One of the most intriguing developments has been phone-in surveys on radio and television programmes. In response to a topical issue, people phone a particular number to indicate a yes or no preference.

These are particularly inexpensive to operate with new technology; they cost very little to initiate and the results are available in real-time. These phone-in surveys provide a low cost and non-binding alternative to referendums; in Ireland, phone-in surveys have had participation rates of over 10,000 phone calls.[8] These high participation polls provide a snapshot of public opinion which policy makers are likely to take seriously.

7.4 International politics

7.4.1 *International pressure groups*

New technologies are also facilitating international pressure groups and interest groups. Because of low 'start-up' costs and low communication costs, technology has been used even more effectively at international than at national level. There is no need for expensive mailings or other propaganda – electronic word of mouth can enable ad-hoc groups with no permanent or rigid structure to mobilise large numbers of people to act in concert in dispersed locations. The June 1999 'Reclaim the City' demonstrations and the World Trade Organisation protests have inspired anti-globalisation demonstrations. In June 1999, protests took place in 17 cities around the world, with protesters recruited and organised via websites. Since then, such protests have become commonplace on anti-globalisation themes, usually timed to coincide with meetings of either the World Trade Organisation or the G7 group of nations. It is virtually impossible to find, much less prosecute, the organisers of such demonstrations, and even less possible to stop co-ordination amongst interested parties of such protest activities. Such informal groupings seem hardly worthy of the term 'organisation': they lack a formal structure and are recreated afresh at each new event. They can appear and disappear as quickly as newsgroups and discussion lists in 'cyberspace', with temporary websites and discussion lists which lapse once the protest is finished and with new sites and lists circulating, at virtual speeds, when a new protest is being organised. Using text messaging, individuals can also rapidly congregate in specific areas and just as rapidly disperse. These groups, however ephemeral, have 'real space' manifestations in concrete political actions, which disrupt activities and claim headlines, and are mobilising people across nations (see Surman and Reilly 2003).

These groups may start as informal associations, but some become formal and even legitimate. Ecology and nuclear disarmament movements, already mentioned in the context of national politics, are both examples of groups whose members are dispersed throughout the world and would have little face-to-face contact with each other. New ICTs have been used to develop common policy statements and co-ordinate protest activities. ICTs also permit public relations (or propaganda) activities that attract new members to the group through an enhanced public profile. GreenNet, PeaceNet and ConflictNet all developed as means of articulating specific international policy objectives, but have become permanent and well-organised interest groups. They joined together to create the Association for Progressive Communications (APC) in 1990 (Frederick 1993); the APC includes groups located in the former Soviet Union, the Pacific Rim, South America and Africa, and so provide an umbrella for an truly international pressure group.

The actual political impact of such groups is sometimes debated, but there are instances where their actions are claimed to have had significant impact. One example has been Amnesty International's use of the Internet to organise protest letters when human rights violations are uncovered. It is claimed that hundreds of political prisoners have been freed as a result of such co-ordinated pressure campaigns (Odasz 1995:119). Equally, NGOs have been very vocal in international environmental conferences, exerting great pressure on governments and proclaiming policy statements that governments agree to at such conferences. In recent international environmental and women's conferences, these voluntary organisations have often played a very prominent public role in deliberations. In the case of the Women's International Congress in China in 1996, it has been suggested that the denial of access and participation by some women's groups became an issue through electronic discussions and protests, then became an issue at the conference, and resulted in a UN working group being established to examine the issue. In some instances, therefore, the activities of these dispersed electronic groups have resulted in policy outputs.

7.4.2 *International expressions of internal dissent*

An additional use of new technology is the international expression and organisation of dissent in relation to one country. In some parts of the world, dissident groups cannot legally express their views within the country and, even with new technology, the development of alternative political groupings is suppressed. Formerly, their only propaganda outlet would have been underground press publications or offshore radio broadcasts. Now, individuals can convey their views to an international audience using new technologies, and can also use 'offshore' electronic sites to disseminate information within their own country. Electronic discussion groups on the Internet are often used as fora for opposition to national governments, such as the People's Republic of China and the government of Burma. Such groups are also a means by which emigrants can keep in contact with each other, for social support as well as political action. Inexpensive video cameras, and tape recorders have all become means of capturing information and, especially if the visual or audio information is digitised, making it available to a wider audience, including transnational news organisations such as CNN. World Wide Web sites become quasi-official sources of information to counter 'propaganda' claims of repressive regimes. With the proliferation of satellite broadcasting and Internet video and audio streaming, it is becoming difficult to prevent such propaganda from being received within the target country.[9] Telephone connections can be used to communicate digitised text, graphics, video information; even verbal information, when rapidly disseminated, has an important impact.

The threat of such electronic dissident action is evident from the extremes to which countries go to suppress it. The publicity that resulted from the suppression of dissent in Tiananmen Square was effective precisely because it was disseminated on an international scale, and on the tenth anniversary of Tiananmen Square, the Chinese government reduced communication as much as possible from citizens to outside journalists and also blocked the reception of foreign news stations. In the 1991 abortive Russian counter-revolution, Yeltsin and his supporters were surrounded in government buildings by tanks and soldiers but were still able to get news of the threat to the outside world, and so organise world support against the counter-revolutionaries (see Castells 1998: 60–2). The difficulty of regulating international information flows is highlighted by the extent to which individuals can have access to satellite broadcasts regardless of attempts by government to restrict information (for an example from Romania, see Gross 1996).

This extra-national discussion now extends to warfare, where conflicting views and reports about conflict are expressed on discussion lists. In the early days, these discussion lists were on Usenet, but are now often mirrored on World Wide Web sites or even hosted on sites where anyone can post messages to be read anywhere in the world. Since the World Wide Web is available to people with little computer expertise, these discussion lists now reach a significantly wide audience. Ethnic conflict in Yugoslavia, for instance, was waged electronically (Taylor 1999), with conflicting claims (often supported by visual images) commonly found on different websites. However, as with the 'fringe' political action groups, the impact of such discussions on policy formation is hard to gauge and may only be as significant as graffiti or handing out propaganda sheets on street corners. The benefit may largely be to facilitate the co-ordination of propaganda activities and the dissemination of information to activists.

It should also be noted that not all activities by emigrants are necessarily illegal. With the increase in transnational labour flows and in access to technology, there are many 'long distance nationalists'. These are people who remain committed to political participation in their home country, whether their residence in another country is temporary or long term. In countries that permit absentee voting (such as the United States), these individuals can be an important political voice. Even in countries where citizens who reside outside the country cannot vote (such as Ireland), committed citizens can make their opinions known via electronic mail or postal campaigns, and can organise themselves as a pressure group. It is notable that recent political history in Ireland has begun to emphasise citizenship and identity as a non-territorial characteristic. This process first gained prominence when Mary Robinson was President of Ireland and specifically discussed the Irish Diaspora. In her address to the Houses of the Oireachtas on 2 February 1995, she highlighted the transformation which new technology was bringing about in the way that those who had left Ireland continued to maintain their Irish identity and their contact with people and events in Ireland (Address by Uachtarán na hÉireann Mary Robinson to Joint Sitting of the Houses of the Oireachtas, http://www.gov.ie/oireachtas/Addresses/02Feb1995.htm):

> We are at the centre of an adventure in human information and communication greater than any other since the invention of the printing press. We will see our lives changed by that. We still have time to influence the process and I am glad to see that we in Ireland are doing this. In some cases this may merely involve drawing attention to what already exists. The entire Radio 1 service of RTÉ is now transmitted live over most of Europe on the Astra satellite. In North America we have a presence through the Galaxy satellite. There are several Internet providers in Ireland and bulletin boards with community databases throughout the island. The magic of E-mail surmounts time and distance and cost. And the splendid and relatively recent technology of the World Wide Web means that local energies and powerful opportunities of access are being made available on the information highway.
>
> The shadow of departure will never be lifted. The grief of seeing a child or other family member leave Ireland will always remain sharp and the absence will never be easy to bear. But we can make their lives easier if we use this new technology to bring the news from home. As a people, we are proud of our storytelling, our literature, our theatre, our ability to improvise with words. And there is a temptation to think that we put that at risk if we espouse these new forms of communication. In fact we can profoundly enrich the method of contact by the means of expression, and we can and should – as a people who have a painful historic experience of silence and absence – welcome and use the noise, the excitement, the speed of contact and the sheer exuberance of these new forms.

A recent change in the Irish constitution redefined national identity on a less territorial basis, in order to accommodate the Good Friday Agreement in Northern Ireland. This is part of a continuing process in which those who are no longer resident in the state continue to identify with it and wish to have a political input in its policy decisions. New technologies allow individuals to participate in national or global politics, regardless of where they reside.

7.4.3 *Illegal politics*

Some political groups engage in activities that are deemed to be illegal or subversive in particular countries, but are accepted elsewhere (e.g. protest movements in China or Burma). But there are also groups whose policy aims are illegal almost everywhere. These would include white supremacist, racist and anti-Semitic groups as well as terrorist groups. The World Wide Web, discussion lists, and electronic mail are used by a large number of such groups (Whine 1997; Castells 1997). The use of the Internet for such illegal activities is virtually impossible to monitor, much less regulate or stop. It is possible, as we have seen, to set up a website in one jurisdiction that promotes activities that are illegal in another, and it is very difficult to prevent access to it.

In addition, there may be countries that actually disagree about what is illegal. There is a great legal diversity in the world, and United Nations conventions that promote harmony tend to be vague about specifics. Finally, there are some jurisdictions where such laws may actually be impossible to implement. In the United States, the First Amendment protection for free speech makes it very difficult to prohibit sites discussing activities that many other countries would deem illegal.

If it is difficult to monitor or control the sites that are the centre of illegal activities, it is equally difficult to find those who access them. Tracking people who access such sites is a relatively complicated and expensive process. Do such illegal electronic activities have sufficient impact to warrant such concern? As with all organisations, communication within illegal organisations has two functions: internal co-ordination and external propaganda. In terms of external propaganda, it is 'public' distribution of propaganda that is illegal; private distribution through personal conversations is usually not illegal and arouses little comment. Racist websites are illegal in the same way as handing out pamphlets at a street corner is illegal, because of their public character. Such websites are worrying because they are so hard to control, but there is little evidence as of yet that such sites have significant influence or power. Users of the Internet still have to seek such sites out, just as they would previously have had to write away for such information. There are differences, of course – the 'address' is easier to find, and the distribution of the information is less costly – so the cost of such propaganda is less. However, there is still no strong evidence that using the Internet makes these groups more effective than when they were dependent on more traditional propaganda.

The internal co-ordination function of such groups is a different issue, however. Especially when encryption technology is used, illegal groups now have new resources to communicate and co-ordinate activities amongst members. Individuals can exchange addresses and can use websites with password access, they can use bulletin boards that are accessed via telephone rather than Internet, and they can use anonymous sites that enable users to upload or download files. These actions are difficult to detect, although it is possible to monitor the actions of known participants through surveillance technology. Yet this is nothing new. Previously, individuals co-ordinated activities through phone calls, which the state could monitor only when it first knew which individuals to monitor. Similarly, the state can monitor email or web activities once the individuals have first been identified. The problem remains initial detection. What are the various email addresses being used? What mobile phone numbers are being used (especially in these days of pay-as-you-go mobile phones in which the user is anonymous)? This is part of a long history of clandestine communication among illegal groups, and a problem for police and criminal investigation.

The one new element is the question of encryption. Individuals have often, in the past, wished to communicate in ways that permit secure communication. There is a long history of using codes to disguise the meaning of letters or other communications so that the communication, if intercepted, cannot be deciphered. In more recent times, phone scramblers have been used to ensure secure communication, even if phones were tapped. However, such codes are often difficult to keep secure and cumbersome to use. The real threat of new communication technologies is the development of secure encryption systems that makes it possible for anyone, with only a small degree of knowledge, to communicate in a secure manner. Indeed, governments, even if they can intercept the communication, may still be unable to decipher the content. At the moment, though, it is more a worry than a reality. There was concern, for instance, after the events of 11 September 2001, that terrorists were using secure encryption to organise their activities (Campbell 2001). In fact, since so few communications are currently encrypted, an encrypted communication would be very easy to discover. It is more revealing to encrypt a message than simply to use plain electronic mail, because it identifies people whose activities should then be monitored.

7.5 Irish political participation

Ireland can illustrate the potential of new technology for political change. Ireland is not a unique political system, but new technologies will have a particular impact on the Irish political landscape. Although Ireland is a parliamentary democracy with a professional civil service, it is also a post-colonial society (for a summary, see Coakley and Gallagher 1999; Chubb 1992; Peillon 1982; Clancy, Drudy et al. 1986). There was an expectation that government benefits were not allocated on the basis of objective criteria, but as a result of the personal assistance of politicians. This type of politics has been described as clientelist: politicians intercede on behalf of voters and, in return, voters become clients, providing votes at election time in reward for benefits (Komito 1984). Such a political system is not uncommon in post-colonial nations, where the administration of the state becomes politicised and personalised (Clapham 1982). In fact, the system actually operated relatively fairly, but in a closed and non-transparent manner. The 'influence' of politicians derived from their knowledge of how the system operated, the unwillingness of Irish civil servants to respond directly to citizens, and long delays in providing benefits (especially prevalent during the expansion of state activities in the 1970s and 1980s). Citizens who experienced long delays in getting state benefits, with no adequate feedback from civil servants, were dependent on any assistance that politicians could provide. This tacit exchange of political support for special personal preference has been a cornerstone of Irish politics since independence in the 1920s (Komito and Gallagher 1999).

This political relationship between citizen, politician and the state has changed as a result of information technology in the government over the past twenty years. With the increase in state intervention in Ireland in the 1960s came an increase in the amount of work in government departments. Office information systems were eventually introduced in the Irish civil service, although it was not until the 1980s that a dramatic increase in IT-related expenditure, in both equipment and staff, took place (Pye 1992). The justification for IT investment was to improve the efficiency of service provision, and indeed the speed of processing cases increased, as did the ability of civil servants to deal with more complex eligibility criteria, using ever more information about applicants. As a consequence, however, the basis for politicians' special influence was undermined. Previously, people needed to monitor the progress of applications (Komito 1989), and politicians 'sold' their ability to provide this information. With the introduction of office information systems, the processing of cases speeded up, and the need for intervention to discover the status of a case lessened. Furthermore, the direct monitoring of cases by applicants became possible. Under the former system, it was difficult on a practical level to find out exactly what was happening with a particular case. The answer might be found only on a particular piece of paper on a particular desk, and it might not be clear on whose desk the case was, and, if the person was away or busy, a report might be slow in coming. Direct queries by citizens produced either no answer or an answer only very slowly, but civil servants had little choice but to put other work aside to find the answer if a politician enquired. With new technology, a departmental information officer (or even a receptionist) can now easily trace the progress of cases.[10] Citizens no longer need politicians, they need only to post a letter or make a phone call (often free of charge) to the relevant department to monitor an application's progress (Komito 1997). Local government authorities have been investing in new technologies which improve internal administration and the delivery of services, but which also enable direct input to be made by individuals and groups. Increasingly, if an individual or group wants to comment about a proposal or find out current state of services, they can contact the local county council directly.

No longer are citizens dependent on the mediation and intervention of politicians for information. Politics in Ireland had previously been conditioned by restricted access to information, but this has now changed, as an unintentional consequence of efficiency-driven ICT investments. This has altered the nature of the relation between citizen and state in Ireland. Individual issues are less central to the political process, since politicians do not need to be involved in the provision of individual benefits. This leaves scope for the development of collective policy inputs, and perhaps it is no coincidence that the 1990s has seen a substantial growth in policy based political parties (Democratic Left, Progressive Democrats, Green Party), as well as 'fringe' groups acting to advance local community interests. These political changes have also lessened the dependence of individuals on others to mediate their access to power, whether that mediator was a politician or other influential figure. This lessened sense of dependency allows citizens to have a more active voice in the determination of public policy in Ireland. Of course one should not overestimate this impact; by and large, citizens still acquiesce in decisions being made by influential figures in Irish political life. However, it is hard to ignore the changes that have taken place of late. Technology has reduced the barriers for participation; it is easier for local groups to develop and their activities can be more efficient and effective.[11]

Not only has there been a growth in policy-driven political parties, but there has also been an increase in transparency and accountability. This has been demonstrated by numerous tribunals investigating the corruption that had been widespread in the 1980s and 1990s, and can be seen as a consequence of direct access to information under Freedom of Information legislation. The continuation of clientelist politics in Ireland depended on restricted information, when private decisions regarding public policies were never subjected to scrutiny. Direct access to information reduces dependency on politicians for assistance, enabling individuals to have all the relevant information themselves. Freedom of Information legislation ensures access, not only to personal information, but to all government information. Such public access to the decision process exposes decisions made privately, potentially for private benefit. The result is more explicit decisions which are removed from the aura of backrooms and private benefit for the well connected (see Komito 1999).

Chapter 8

State policies and the information society

Changes in politics and social relations are inevitable – even if unintended or unplanned – consequences of technological and economic change. One of the common themes in the information society is that new technology is making distance irrelevant, and where people live and work is changing, as a result of new technology. This is because communication costs have reduced, while the kinds of information that can be communicated have increased, so that location is not longer a limiting factor in economic growth or everyday life. Since these changes will have an impact on how people live and participate in society, such changes should be directed by public policy and may require policy decisions (for instance, regarding the provision of the appropriate infrastructure). The danger is that governments will have no policy, and so economic and technological forces will be responsible for determining the future political, social and cultural life of nations. Since new technologies may be capable of radically altering the nature of the state and citizenship, and even the very nature of the state, a 'laissez-faire' policy attitude towards the information society is not sustainable.

8.1 Information society policy

Ever since the development of cheaper computing in the 1970s, and especially with personal computers in the 1980s, there has been a general perception, articulated by Bell (1973), Masuda (1981) and Nora and Minc (1980) amongst others, that future economic success depends on new technologies. This was echoed in more popular writings by people such as Toffler (1980), Stonier (1983), and Tapscott and Caston (1993). The result was the development of policies regarding the role of governments in this ongoing technological change. The United States in the 1990s, for instance, formulated a policy to provide a 'National Information Infrastructure' (otherwise known as the Information Superhighway). This policy was articulated by Senator, subsequently Vice-President, Al Gore, in 1991: 'as a nation we will invest in the critical infrastructure of information superhighways' (Gore 1991); he argued that just as road traffic and interstate commerce increased once a national road system was built, so would data traffic and electronic commerce increase if high-speed networks were available. Underlying this proposal was the common assumption that nations were in economic competition: 'The alternative is to wait until other nations show us how to take advantage of this technology – and they will. We must move first.' The United States may have been first with the phrase 'information superhighway', but was by no means first to suggest that government policy needed to respond to technological change. In the United Kingdom, the then Prime Minister James Callaghan referred in 1978 to the 'microelectronics revolution', commenting that 'we must prepare for it . . . [if we are] to reap the maximum benefit for the new technology'; this was followed by a host of UK government initiatives (Robins and Webster 1999: 64–5), including the Information Society Initiative in 1995, the National Strategy for Local e-Government, and, most recently, the Office of the e-Envoy (www.e-envoy.gov.uk), whose aim is to 'get the UK online, to ensure that the country, its citizens and its businesses derive maximum benefit from the knowledge economy'. To support this aim, the Office has four core objectives:

- to make the UK the best place in the world for e-commerce
- to ensure that everyone who wants to can access the Internet by 2005
- to deliver electronically, and in a customer-focused way, all government services by 2005
- to co-ordinate the UK government's e-agenda across different departments

The equivalent European document was the Bangemann Report (Commission of the European Community 1994), which also articulated a 'free-market' approach to telecommunications infrastructure (for a comparative discussion of US, UK and EU positions, see Dutton, Blumer et al. 1996).

Virtually all states have e-Government policies with somewhat similar aspirations; they all tend to focus on the use of technology to enhance economic competitiveness, and the use of technology to improve the efficiency of services. The outcomes of these policies can be measured relatively easily, and such surveys appear with great regularity, with countries rated differently on electronic government and public access to the Internet (Organisation for Economic Co-operation and Development 2003c; 2002; International Telecommunication Union and Minges 2003; Organisation for Economic Co-operation and Development 2003b; Commission of the European Community 2000). Government programmes vary on two dimensions. One is the percentage of the annual budget which a country has decided to invest in such technology programmes, although governments try to spend as little as possible on modernisation programmes. The more important dimension of variation is the extent of government support for programmes which have a social component. For instance, both the United Kingdom and South Korea similarly want their citizens to have high-speed Internet access (i.e., broadband). However, while the United Kingdom hopes that private investment will provide this access, the South Korean government supports the infrastructure and is paying one half of the cost of upgrading the network to provide even higher speeds.[1] This variation exists even within the European Union. In Danish and Finnish policy, the social component was an integral part of the national plan (see Friis 1997; Castells and Himanen 2002); in other cases, the social component was seen as an additional benefit that would somehow accrue to societies once the infrastructure was in place. For instance, the Bangemann report expected that the information society would lead to 'more efficient, transparent and responsive public services, closer to the citizen and at lower cost', and, two years later, 'Building the European Information Society for us all: First Reflections of the High Level Group of Experts' (Commission of the European Community 1996) noted that 'ICTs create new opportunities for greater public participation in and awareness of the political process'. In the United States, the 'National Information Infrastructure: Agenda for Action' (Information Infrastructure Task Force 1993) noted that

> All Americans have a stake in the construction of an advanced National Information Infrastructure (NII), a seamless web of communications networks, computers, databases, and consumer electronics that will put vast amounts of information at users' fingertips. Development of the NII can help unleash an information revolution that will change forever the way people live, work, and interact with each other.

The result of these policy initiatives has been some investment in infrastructure and some support for those who cannot afford to participate in the technological revolution, but governments primarily encourage developments and investments in the private sector through legislation and regulation. In the case of the European Community, the free-market could be achieved only by opening the national telecommunications companies to private competition, thus leading to a decrease in telecommunications costs. This has, by and large, been successful: while telecommunications costs are not yet at the level of the United States, they are now far lower, with far greater flexibility than before. However, social and political benefits of the technological revolution have been slower to develop. Benefits have been largely in the realm of administrative efficiency, so that citizens have easier access to services and benefits from governments.

8.2 Irish information society policy

Policy regarding the information society in Ireland is located in the Department of the Taoiseach (Prime Minister) and is articulated through the Information Society Policy Unit and the Information Society Commission, with the first public statements coming from the government appointed Information Society Steering Committee (1996). Investment policy in Ireland has focused on e-government (enhancing delivery of services) rather than e-democracy (enhancing political and social participation). For instance, *New Connections*, the second Government Action Plan on the Information Society, launched in April 2002 (Information Society Commission 2002), focused on the policy strands of infrastructure (telecommunication infrastructure, legal and regulatory issues, eGovernment) and support frameworks (eBusiness, research and development, lifelong learning, and eInclusion). Progress on infrastructure has been uneven, but with some successful projects. For instance, submission of tax returns and payment of income tax through the Revenue Commission can now take place electronically. In 2002, 23,000 self-employed taxpayers filed electronically (nine per cent of total self-employed that year); in 2003, the number rose to 100,000 (estimated to be 40 per cent of self-employed).[2] There are many other projects in progress, such as payment of car tax, registration of births and obtaining birth certificates, and access to planning applications.

One of the most significant developments in Irish policy has been a move away from dependence on the private sector to provide high-speed Internet access for citizens. Current policy is directed towards co-funding construction by local and regional authorities of fibre optic networks around the country which would facilitate broadband access. The resulting networks would be an open access broadband infrastructure, which would remain in public ownership, but would allow access by private operators. Another example of government investment has been the Digital Media District in Dublin. The Digital Hub (www.thedigitalhub.com) will provide a 'nursery' for start-up digital industries, and the government is investing in providing a high-speed fibre optic network. The Hub developed out of the establishment of Media Lab Europe, an offshoot of the MIT Media Lab. Established in 2000, Media Lab Europe has received government funding, as well as investment from national and multinational corporations. It is hoped that such a research and development facility will encourage the development of commercial applications of new technologies and inventions.

It is notable that reports of government activities recount specific policy achievements regarding infrastructure, but reports on issues such as eInclusion and lifelong learning focus on pilot projects and the need for reviews rather than on concrete achievements. Investment for service delivery and economic development may have been forthcoming, but government investment in public access has been slower. Even the infrastructure investments benefit companies more than individuals, since a high-speed network is of little assistance to people who cannot connect to the network. Unlike countries such as Denmark and Finland, Ireland has focused on encouraging private investment in infrastructure and encouraging competition to keep prices down. Although some public Internet access is provided through public libraries, the assumption remains that individuals will have access to the Internet from home by privately paying the cost of computer ownership and telecommunications access (both of which the government hopes will remain low owing to market competition).

The Irish government expectation that individuals will gain access to the Internet by buying their own computers has been partially borne out. According to the Central Statistics Office (2003b), the percentage of households with computers rose from 18.6 per cent in 1998 to 42.3 per cent in 2003 and, more significantly, the percentage of computers owners who have Internet access rose from 26.7 per cent to 79.6 per cent in the same period. This statistic might be a bit misleading, since it is now common practice for computers to have Internet capabilities preinstalled. However, other statistics suggest that these capabilities are used; 37.6 per cent of individuals have actually used the Internet and almost 25 per cent use the Internet at least once a week. Individuals use the Internet to communicate with others (83.5 per cent of Internet users) or to find information (84.5 per cent), but only 38.5 per cent use it for either government information or services. In accord with usage patterns elsewhere, gender is no longer an issue, as women are as likely to use the Internet as men. However, age, occupational status and geographical location continue to be relevant; the least likely individuals to use the Internet are those over 55 (with a noticeable number between 45 and 55 who do not use it), those residing in least economically developed regions of the Midlands, West and Border areas, and those who do not work in offices. However, these statistics also focus attention on the 63 per cent who have never used the Internet and the additional 12 per cent who use it rarely. Even more significantly, 58 per cent of households do not even own a computer. The number of Internet users has dramatically increased, but there remain many who do not use it, whether on account of economic barriers or lack of perceived benefit. The government policy of minimal intervention in the market place has still left a number of individuals without access to the information now available through eGovernment reforms.

8.3 Governance and the 'digital divide'

Exclusion and differential access to new technologies are significant policy issues because many countries have become concerned about increasing political apathy and decreasing voter turnout in recent decades. Fewer people vote in elections and political parties report difficulties recruiting younger activists. There has been a move from party politics to community and single-issue politics and candidates, indicating a distrust of the traditional political structure. New technologies have led to a growth in the size and complexity of administrative structures, which appear more distant from and less accountable to the voter. Such diminished political participation is indicative of apathy and alienation. This means a loss of legitimacy for government, and governance without legitimacy is problematic. Thus, in the European Union, there has been much talk of the 'democratic deficit', and similar concerns about voter apathy exist in the United States. Governments accept the need for positive policy action to make government more open and transparent in an attempt to increase participation and accountability. The alternatives are an electorate which acquiesces because citizens are distracted by 'bread and circuses' or are forced to acquiesce by coercive sanctions. Few would argue for the desirability of either option.

Many government policy documents suggest that new technologies will improve public participation and so reverse this trend. However, new technologies can only provide possibilities. For instance, The High Level Expert Group on the Social and Societal Aspects of the Information Society (Commission of the European Community 1996) noted that 'ICTs create new opportunities for greater public participation in and awareness of the political process', but also warned that 'the increase in the flow of information does not necessarily engender an amelioration of the democratic system. It could just as easily lead to a distancing of citizens with regard to real democratic stakes'. It is the policy decisions of governments that will determine whether new technologies foster greater participation or provide additional barriers between citizens and states. The exclusion of citizens from access to new technologies would prevent such participation. Policies for 'Electronic Government' or 'Open Government' are of little benefit if people do not have access to the necessary communications technology. Public information that is not accessible is hardly 'public'. If people need information about government, then providing access to such information is part of satisfying that need.

Statistics such as those given on the previous page suggest that policy decisions in Ireland have not so far had the desired outcome: many people do not own computers, many of those who do see no benefit in using the Internet, and those who use the Internet do not access government services and information. Ireland is by no means unique in this; effectiveness is usually related to the amount of money which governments are willing to invest in public access technologies. If governments wish to improve access to information, then they have to provide support for individuals who purchase computers by providing grants or tax relief. Since computers are used for so many other purposes (especially entertainment), governments are often reluctant to provide financial support. They assume that the other benefits of home computer ownership will provide sufficient incentive for computer purchase.

A related problem is the question of telecommunications charges. Even when people can afford a computer, many are deterred by the charges of using a telephone line or high-speed data line in order to access the Internet. In some countries, this issue does not arise, as there is a flat rate charge for all local telephone use. But in countries where even local telephone calls are charged individually, cost is a deterrent to use. This issue has been faced before. When postal services were introduced, a policy decision was made that it should cost everyone the same to post a letter, regardless of the location of sender or receiver. In fact, the true economic cost of posting a letter varies, depending on location. Such services are more expensive to operate in low-density rural areas, and so the actual economic cost of posting a letter from one rural area to another rural area is much higher than posting a letter from one urban area to another. Despite this, the cost to the consumer is the same, and the higher economic costs of posting rural letters are shared by all. Similar issues arise in the provision of telephone services and transportation services such as buses and trains. In some cases, the cost remains the same, with cheaper service locations subsidising more expensive service locations. In other cases, the government subsidises the cost to all consumers because access to such communication and transportation services is seen as a right for all citizens. Similar decisions have to be made regarding telecommunications charges: should governments ensure that everyone can afford access to the Internet by subsidising the cost?

This still leaves the question of individuals who cannot afford to purchase a computer. Should governments provide inexpensive public Internet access for those who cannot afford the cost of buying and using a home computer? Enabling access to information is also not a new problem. As books became an important means of distributing information, the problem arose of those who could not read and could not afford to buy books. The dual solutions were to fund literacy programmes and, more importantly, to create public libraries. The public library is a revolutionary idea: those who cannot afford to buy books have the entitlement to borrow a book, thus ensuring that everyone has access to information. Should there be a similar policy regarding electronic information? Should those who cannot afford to buy a computer have the right to 'borrow' a computer, or at least borrow access to a computer? In many cases, this right is exercised in the same way as the right to books, and public libraries make it possible for people to 'borrow' computer time just as they previously borrowed books.

Current policy in Ireland, according to a government action plan (Information Society Ireland 1999), is to provide high-speed Internet access in all public libraries, and further consideration will be given to 'extending access to those who do not have PC/Internet access at present' (p. 7), including dedicated kiosks and use of schools and post offices. The option of subsidising telecommunications costs or training costs tends to be underplayed; government policy is to provide access in public locations, rather than facilitating people to have access at home. This is rather like providing public telephones, without enabling people to have telephones in their own homes. Will political participation and access to government services increase if people have access only at specified times and places in public settings? The more effective, though more expensive, policy solution would be for governments to subsidise the purchase and use of computers in the home.

Should governments also ensure that individuals have access to the necessary technology, and training in how to use that technology? Some services involve 'stand-alone' competencies (such as learning how to use an automatic teller machine, or social welfare users learning to use a smart card that enables them to claim benefits) that do not necessarily transfer to other services. However, many of the services, whether information about government policies, or individual benefits delivered to citizens, are based on the Internet because the Internet is seen as the most widely available and flexible communication technology. When used with a World Wide Web interface, it is seen as easy to use by citizens while permitting a multitude of services to be provided in the one format. Will governments ensure that citizens learn how to use computers as well as the Internet?

In addition to owning or accessing a computer, affording the cost of telecommunications access, and possessing the skills to use computers, there still remain other barriers to bridging the 'digital divide'. The statistics in the previous section demonstrated that a large number of people do not use the Internet because they do not see it as useful, and certainly do not see government-provided information as useful. Information has to be made available in the appropriate format: citizens have to be able to find it and understand it. The earliest attempts by governments in the 1990s to make information available to the general public tended to be ineffective. Information was transferred from paper form to electronic form without changing the information or its format. Information was organised on government websites according to categories and using descriptions that officials were used to but which the public found impossible to understand. The information was available but no one could find it or understand it once found, so it was meaningless.

Fortunately, many governments now organise information in terms that make sense to individuals. For instance, a government website may first distinguish between information needs of individuals versus organisations, and then, within the category of individual, may organise information according to the life events that are likely to have brought individuals to the website in the first place (birth, death, marriage, taxes).[3] These changes at least mean that people can find the information and that it will be provided in a meaningful way. Governments then have to engage in publicity campaigns to demonstrate the value of such information to people who clearly do not see any particular benefit to it.

Many of the policies promoted by governments (including the Irish government, see O'Donnell, McQuillan et al. 2003) tend to assume that the benefits of new technologies are obvious and that the only problem is to demonstrate the benefits. The evidence for such an assumption is tenuous. For instance, in the case of the Information Age Town project in Ennis, County Clare, individuals were given computers, inexpensive telecommunications access and free training in the use of computers. After three years of this encouragement, 25 per cent did not want a computer and, of those who had computers, only 35 per cent accessed the Internet more than once or twice a week and nearly twenty per cent did not access the Internet at all (Behaviour and Attitudes Marketing Research 2001). This is not atypical, as recent research in the United Kingdom revealed similar patterns (Wyatt, Thomas et al. 2002). While barriers to participation in the digital revolution exist and are often not addressed by government policies in any effective way, it is worth remembering that, even in the absence of the barriers previously discussed, individuals might simply view the digital revolution as irrelevant in their lives.

It is arguable whether this state of affairs will last very long; people may well find that they have little choice but to participate in the digital revolution. Increasingly, information is either available only electronically, or else it is inconvenient or expensive to access in non-electronic form. Thus people may find themselves banking online simply because the local bank branch has closed. They may find themselves checking for information online because only summary information is available via radio or television. As electronic access becomes mandatory rather than optional, barriers to participation become even more important. If these barriers are not effectively addressed (which requires more than just publicity campaigns), many people will continue to be excluded from participation.

8.4 New technology and development

Many governments, including the Irish government, assume that distance is no longer an economic barrier in the digital economy. As long as there is an inexpensive and fast global information infrastructure, peripheral countries can participate in a world economy as effectively as countries that are physically close to the economic core. If true, this has important implications for governments in their internal development as well as national development policies, since it would follow that rural areas of a country could participate as effectively as urban areas. In economic life, the three most important issues have traditionally been location, location, location; many governments hope that new technologies will make that mantra obsolete.

There is some evidence to support this view. Reduced telecommunications costs mean that companies can take advantage of low labour costs in different parts of the world. Many multinationals have created 'back-office' operations in low wage locations, where employees carry out data entry or other activities for lower wages than multinationals would have to pay elsewhere; this has been an effective strategy for attracting some types of industry to Ireland (Forfas 2002; 1999). There are many factors that can determine the profits to be made from such back-office operations. There is the cost of training the appropriate labour force with the necessary skills, while also maintaining low labour costs. One way to do this is for governments rather than individuals to underwrite training costs. However, in such a situation, will people with similar skills (computer inputting, English speaking) but in poorer areas (India, Asia, Caribbean) be able to undercut wages?[4] Wage costs are not the sole determinant of profitable investment. There are also costs for the communications infrastructure; if the communications costs are too high or connections too unreliable, then low wage costs would not be sufficient to make the investment profitable. If governments provide the necessary infrastructure, then secure and reliable high-speed communication can allow a location to remain competitive, even if labour costs or training costs increase.

Irish governments have been following this policy of absorbing telecommunications costs and training costs of workers, so that the combination of reduced labour costs and inexpensive high-speed telecommunications will make investment attractive to multinational corporations. This strategy appears to correctly gauge the criteria used by multinationals who do not just look for cheap labour when deciding on investment. They look for trained labour, and labour which can adapt and learn new skills as necessary. They also want backup, in the sense of a broader skills pool to provide support, and government strategies to adjust training needs as industry needs change. This includes providing education courses in Institutes of Technology (formerly

Regional Technical Colleges) which suit particular industry needs (Wickham 1998a), as well as providing high-speed networks that would link local offices with locations elsewhere in the world (Lillington 2000). This policy seems to have successfully encouraged inward investment by multinationals and the development of an indigenous software industry (Ó Riain 1997; 1998; 2000; see also Lillington 2003; McCaffrey, U. 2003; Taylor 2003). In the 2004 budget, the Irish Government introduced tax relief for companies investing in research and development, which would make it attractive for foreign companies to locate research activities in Ireland.

Irish government policy regarding rural development was articulated by the publication of a National Spatial Strategy (Government of Ireland 2002b). This policy accepts that there are limits to dispersed economic development; it is not possible for individuals or small groups in isolation to compete effectively in the global economy. It is necessary for there to be a critical mass of individuals and organisations, and for technological (and other) investment to identify geographical areas of potential growth and focus infrastructure investment in those areas. On the other hand, government policy also assumes that once a critical mass of individuals, organisations and infra-structural support is available, these centres of growth would utilise new technologies to compete effectively in the global economy. Thus the provision of high-speed telecommunications in rural areas will encourage the development of growth centres. In December 2003, the government announced an ambitious programme of decentralisation, in which entire government departments were to be relocated outside of the capital to regional locations. It is too early to know whether the programme will actually be carried out or whether the relocation will lead to efficient administration, but even the prospect of moving ten thousand administrative jobs outside the administrative centre is possible only as a result of new technologies of communication and information processing.[5]

Rural policy is intended to achieve social as well as economic objectives. Providing telecommunications infrastructure can improve both the social and economic viability of rural communities by giving individuals and organisations access to world markets at minimal economic cost. Governments find that individuals are migrating towards cities and out of rural areas because of the economic opportunities in urban areas. Many governments want to maintain viability of rural communities, and solve the social and demographic problems of rural depopulation. In some cases, it is as pragmatic as avoiding further congestion of urban areas and also reducing employment costs (since the cost of living in rural areas is less than in urban areas). Rural areas are often considered the 'real' embodiment of the society, so it is important to maintain them. It is expensive to maintain rural communities through state payments to individuals; better to encourage employment growth through education and technology investment. A trained workforce will make the area an attractive location for external investment, since it will enable companies to employ skilled workers at lower salaries. From the worker's perspective, instead of people leaving rural areas to go where the work is, the work will travel, electronically, to where the workers are. There are numerous examples of individual companies that have used new technology to market their commodities successfully, even when those commodities are physical entities that still have to be shipped abroad. Perhaps two of the best-known examples are the Bodhrán makers, Roundstone Musical Instruments (http://www.bodhran.com/) and Kenny's Bookshop & Arts Galleries in Galway (http://www.kennys.ie/). If such economic growth can be encouraged, the social objective of maintaining rural communities can also be achieved.

The use of technology to increase economic participation in global markets is a two-edged sword. While it enables individuals and companies to access global markets, it also enables others to compete with them. Thus an Irish company that sells its software services to a US firm may find itself competing with an Indian company selling similar services, but at a lower cost. Those companies that benefit most from a global information economy are ones which sell a niche product that cannot be easily produced elsewhere (e.g. bodhráns), or ones which are able to still provide the best mix of labour cost, skills and communications.

Despite the rhetoric about the 'death of distance', evidence suggests that location remains relevant even in the new global information economy (Castells and Hall 1994). It may be true that information, as a commodity, is relatively immune from delivery costs associated with agricultural or manufacturing production, but other costs remain significant. The Silicon Valley in California and Route 128 in Boston remain significant because the production of the commodity of information continues to be influenced by geography. People need to talk to each other, often face to face, as there are limits to the efficacy of computer-supported communication systems for exchanges of ideas. They also need to be able to move jobs relatively easily, which supports rapid innovation and adaptation of ideas within an industry. There needs to be a supply of new labour, and there needs to be a technical infrastructure for such production activities to flourish. All of these suggest that innovation tends to cluster; capital investment in a location (including social capital and training facilities) can improve the competitive advantage of one location over another (Castells 1996; Grimes 2000).

Thus, when governments seek to improve the economic position of rural areas, the provision of a distribution infrastructure (high-speed communication lines for domestic use) may increase the consumption of information commodities such as movies on demand or tele-banking, but economic growth also requires a cluster of people, educational opportunities, and other intellectual resources (Ó Riain 1998; Wickham 1998b; Mjøset 1992). The Irish government has to ensure that there are locations which have a critical mass of information activities (to allow for information sharing and employment mobility), the necessary high-speed technological infrastructure to enable the creation and then distribution of such products, and an educational system that produces the appropriate skills base for entry into such employment. Scattering such facilities over a broad area will not produce the critical mass that clearly exists in centres of innovation elsewhere; it must be localised and concentrated.

Although providing a technological infrastructure in rural areas does not ensure economic viability, there may be other justifications for governments providing such an infrastructure. The social benefits of reduced telecommunication costs in rural areas can be significant, as ICTs allow interactions that physical distance would previously have made impossible. It may be too costly to travel into town from an isolated rural area: a car may not be affordable or available, the cost of petrol may be too high, the person may not be able to leave children alone, or may simply not be able to afford the time it takes to drive the long distance. For whatever reason, the result is isolation and lack of access to resources. There is a social loss in the higher opportunity cost in terms of chance encounters. There is a greater chance of interacting with other people, by chance, in a town than outside one's own door in an isolated rural area. ICTs may encourage communication in rural areas, just as telephones did earlier. Governments may provide support for telecommunications infrastructure in rural areas for much the same reasons governments took similar decisions with regard to telephones, electricity, postal mail, television and radio. In these cases, it was decided that it was the right of all citizens to have access to fundamental communications resources, without a penalty based on location.

Universal Service Provision, which ensures 'access to a defined minimum service of specified quality to all users at an affordable price and respecting the principles of universality, equality and continuity',[6] is a policy commitment adhered to by many governments throughout the world and is a clearly articulated policy within the European Union (e.g. European Commission 1996). Such a decision has an economic cost, since providing the necessary infrastructure is more expensive in rural areas, because of the low population density. The higher cost of the rural service is either subsidised by imposing an artificially higher cost on urban consumers, by government subsidy, or by requiring people pay the true economic cost of the service and risk rural dwellers being unable to afford it. Government policies vary on this issue. The European Union policy depends on a combination of free market forces and subsidies. The Commission depends on a liberalised market to keep the cost of services down, but accepts that special subsidies may be required in rural and remote regions, where geographical isolation and low density of population increases the cost of delivery (Commission of the European

Communities 2003). The European Union's optimum solution is tariff rebalancing with targeted schemes for needy users and uneconomic customers rather than general subsidies for access (European Commission 1996). The UK government expects investment to be provided when an economic return on the investment is possible, but encourages competition to ensure that costs remain low. Other countries, such as France, Australia, and Canada subsidise investment in broadband Internet access. The decision is a political one, dictated by policies for rural development as well as those for long-term social and economic benefits. In the case of Ireland, the government is subsidising the cost of a high-speed data network that includes rural areas, but is not subsidising the cost for individual consumers to access it.

8.5 Telework and distance education

Government policies towards rural development are often intertwined with policies for telework and distance education. With new technology, the location of work (or at least some kinds of work) may become independent of the traditional location of 'office' or 'factory'. People can work from home, thus changing the boundary between private/domestic/home and public/work, as well as changing the internal structure of domestic life and relations within the family. Paid employment outside the home is a relatively recent development, linked largely with the industrial development of factories and offices. Before then, agricultural work was carried out on land near home, and skilled craft work could be undertaken anywhere, including at home. Industrial production, however, required power, large machines and shared labour, which meant workers had to come to a central factory location. With the rise of white-collar employment, office work required information (such as files), which had to be in one location, as well as communication with co-workers, which also required face-to-face contact in one location.

New information technologies have changed this to some extent. While industrial production remains tied to a particular location (the requirements of power and machinery remain crucial), much of the production is automated, and remote control of production processes is possible (whether remote entry and editing of newspaper text, or remote monitoring and control of an automated industrial plant). For office work, it is often no longer necessary to physically be in one's place of work. Information in electronic form can be accessed from remote locations and can be shared; electronic communication between office workers can replace face-to-face work. Such information work has become more central, and thus a spatial distinction between home and office is no longer a requirement of employment. Telework is the phrase used to describe working from home, and is often linked with images such as the 'electronic cottage' – allowing the pleasures of rural and idyllic family life while also engaged in well-paid work that is often unavailable in such rural areas. The reality is more complex. After all, cottage industries of earlier years were exploitative sweatshops, populated by overworked, exploited women working by candlelight.

There is a difference between individuals who work for an organisation at home or in tele-cottages (where people simply 'rent' a desk with appropriate computers and telecommunications links) and individuals who have no long-term organisational commitment and sell their services as individual contractors or consultants. In the first case, individuals work at home, but remain full-time employees of a company. Either the company wants to reduce overhead costs by reducing the cost of maintaining office space, or perhaps the individual may otherwise leave employment and the company wants to maintain the person's services. Telework is not the same as 'bringing work home'. For many years, people could bring files home (whether paper or computer), work on them, and then take the results into the office. This is decentralised work, but not telework. 'Telework' implies working outside the office, but having communication facilities so as to interact with individuals and information resources in the same way as if at work. This may involve a high-speed data line, with secure access to company local area network or intranet, or an organisation might operate regional 'drop in' points where employees log on and that desk becomes their 'virtual office' for the duration of log in, or workers might come to a local computer centre and pay for temporary access to a company's network. All of this contrasts with individuals who are self-employed and are selling their services to anyone who will pay. Such individuals either have highly marketable skills (for instance, specialised web design or programming skills) or they are cheaper to employ than in-house workers.

From the societal perspective, telework means that less infrastructure investment in cities is required, with significant reductions in pollution and population density, while permitting rural communities to remain economically viable. People can also work who are restricted to their home or local community for personal or social reasons. Companies can reduce their overheads in, for example, heating and office space (see Forester 1987: 162–4). The advantage for employees is that they can live in nicer or cheaper areas, with reduced commuting time,[7] or work part-time, combining home tasks, such as parenting or caretaking, with paid employment. However, there are problems. Home-workers are less visible in organisations, and are less able to participate in office politics. They may be less likely to be given training opportunities or special jobs that would lead to promotion and so their careers may suffer. Working at home may also be problematic for some people, in that discipline is required to work to one's own timetable, and a particular personality is required not to miss the social contact of an office.

Because telework lends itself to piecework, it can be exploitative. It runs counter to the traditional monitoring mechanism of 'clocking in'. Traditionally, if someone is physically in the building, they are 'at work' and can be observed at their desk or workstation. Many managers are doubtful about telework because it is difficult to monitor the work patterns of home workers. The solution is to monitor work output. That is, if enough output is generated, then the individual must be working at a reasonable pace. It is then tempting to judge the quality of work by the amount of work output, and even more tempting to pay the worker by the amount of work – that is, by piecework. In which case, it is even more tempting to turn that worker into a contract worker, or part-time worker.[8] In a consultancy project, a contract for a specific project can be well paid. But if the work is inputting forms into computers, then payment is calculated per item, which becomes the same as paying per items sold, or sweaters knitted, or policies sold. Such contracts can often lead to low overall salary rates, as they did in non-union factories that paid piecework rates.

In telework, gender can be important: it is often women, rather than men, who want flexible work regimes that enable them to combine domestic and work activities. For many women, telework is a way to carry out domestic tasks and earn money at same time, so as not to have to choose between them.[9] The kinds of tasks that suit such a pattern are routinised tasks that do not require a lot of interaction or communication, and these are often poorly paid. Women have traditionally been paid less than men, so, if telework is female dominated, then as teleworkers get paid less than office workers for the same work, women will continue to dominate a less prestigious, poorer paid sector of the economy (Webster 1996b, see also http://www.emergence.nu).

Of course, there are also well-paid independent contractors who bid for work and obtain or maintain employment while working either at home or in a local telework centre. There are also employees who want to work at home who are well paid. Recent evidence (Central Statistics Office 2003c) suggests that teleworkers in Ireland fall into these categories. A high percentage of teleworkers are managers, administrators or professionals and over 70 per cent have a third level qualification (as opposed to an average of 40 per cent for non-agricultural workers). Telework in Ireland is the preserve of the well paid and well educated. It is also the preserve of men, as two thirds of teleworkers are male. Although most teleworkers are well-paid professionals, clerical and secretarial teleworkers are more likely to be female (76 per cent), while only 28 per cent of managers, administrators and professionals are women. Thus, the notion that telework, gender, and pay are linked seems well founded.

Much of the support for the telework employment model came from governments (national and EU) who had a clear desire to maintain rural populations by providing mechanisms through which rural residents could participate in the labour market without having to leave their locale. The evidence that telework will achieve these policy objectives is ambiguous. The Central Statistics Office reports that in Ireland nearly ten per cent of non-agricultural work is carried out from home to some extent, but only four per cent use a computer with a telecommunications link and only about 2.5 per cent are home-based teleworkers for whom a computer and telecommunications are crucial (Central Statistics Office 2003c). Most strikingly, the greater Dublin region has nearly twice as many teleworkers as a proportion of total workers as do rural regions such as the Midlands and Mid-West. Even for telework, the urban centre continues to be more attractive than rural areas, suggesting that telework may not solve the rural depopulation problem.

This urban bias in telework may be linked to education issues. Those who are selling their skills on the world market first have to obtain, and then maintain, those skills in a competitive environment. Those skills, derived from education and subsequent employment experience, are probably obtained in an urban environment, since centres for innovation with a critical mass of people and resources are still usually urban. Workers must maintain their expertise, and, more importantly, upgrade their skills, since they could become stale and their ability to sell their skills would diminish over time. In an urban environment, workers have a constant interaction with peers and other opportunities to gain exposure to current developments. What about those in a rural setting removed from their peers? How can expertise be maintained? The answer might be through distance education.

Education is crucial to selling one's skills in the new market place. In fact, there has been a fundamental shift in the balance between experience and education. Previously, education provided an initial start, and, during the course of employment, experience provided additional skills and expertise. Employees with longer experience were more valuable than new employees with little experience. However, the new knowledge industries are changing all the time, and previous experience is not necessarily an advantage in adapting to a new technology. New graduates with up-to-date knowledge may have a higher market value than old employees with years of experience but no recent knowledge.[10] However, even recent graduates will find that their expertise may soon be out of date, as organisations, markets, and technologies are subject to rapid change. So, in addition to initial skills levels, all employees, new or old, need to acquire new skills during their working life. Whether they stay in one job or move from one job to another, the relevant skills are now expected to change, thus leading to life-long learning programmes.[11]

With a move to contract employment, training is less and less seen as a company responsibility, but rather as an individual's responsibility to provide for him or herself. The cost of acquiring these skills has to be borne by the individual rather than the company and may have to be acquired in the person's spare time. The state may facilitate the individual since, if governments want an employed workforce, they have to ensure an educated workforce. Daniel Bell appears to be right in that education does appear to be the key to employment in the new global economy. Countries which compete for value-added employment are now focusing on education to create a workforce that can attract high value companies, and thus it is in the state's interest to support people's efforts to acquire additional expertise.

Distance education is not new; the Open University has provided distance education in the United Kingdom for decades, and it has often reached an Irish audience. The National Distance Education Centre of Ireland was established in 1982, but provides only a limited range of undergraduate diplomas and some postgraduate degrees in Information Technology. Throughout the world, the delivery of training and general education using new technologies (email, World Wide Web, and teleconferencing facilities) has increased. There are many obvious advantages to such distance education: instead of going elsewhere to find expertise (and perhaps not being able to afford to travel), remote education gives access to 'best practice' anywhere in world. People who are unable to be in class at specific times because they are working, or have other demands on their time, can still get access to education. The cost of producing such educational packages is reduced, since the same 'product' can be consumed by multiple learners in multiple locations at little extra cost to the producer of the training or education product.

Distance education may be more suited to 'training' than 'education'. While training focuses on specific skills, tasks or competencies, education focuses on ways of thinking and analysis. Computer-based training helps acquire specific skills, but it is not clear how well it works for education. Education is a social process and distance education often lacks satisfactory feedback or interaction with a human teacher, which may diminish the effectiveness of education. Not all knowledge is explicit and objectified; sometimes individuals learn things that they would find difficult to put into words. This 'tacit knowledge' is often conveyed through social interaction, including non-verbal communication, and is an integral part of the education process. Thus it may not be possible to provide it through an electronic medium that tends to be asocial, non-interactive and lacking in non-verbal cues. There have been many attempts to provide faster and more effective interaction, and to use technology to mimic the social environment, but there are limits to its benefits. It has even been argued that training as much as education is dependent on acquiring such tacit knowledge, and this is done through participation in groups whose individuals are practising the skills being acquired (see Orr 1996 for a detailed example of this). These 'communities of practice' (Lave and Wenger 1991; Wenger 1998; Brown and Duguid 2000) suggest limits to the effectiveness of training provided via distance education. The hope that distance education will enable training or education to be decentralised may be misplaced, which must also limit the potential for the decentralisation of work.

8.6 The global digital divide

If information is a commodity, then countries that are impoverished are also unable to afford information. Will this lead to the emergence of 'information poor' societies? For the moment at least, the Internet and the digital revolution, and even the telephone, are features of industrialised societies (Sciadas 2002; World Intellectual Property Organization 2002). Such information poverty is not simply the small number of computers per head of population, but often a lack of entire telecommunications infrastructure. As matters stand, the gap between the information rich and poor is very wide; for example, in a study sponsored by the Canadian International Development Agency, Canada and Finland were measured at 110 and 100 respectively, while India was measured at 5 and Senegal at 7.8 (Sciadas 2002). While the author noted that the gap between digitally developed and underdeveloped was narrowing, it was clear that, without policy intervention, it could be decades before the gap narrowed substantially (Sciadas 2002: 23). Does such a lack of computers and telecommunications actually increase poverty, or is it simply indicative of economic deprivation, which investment in ICTs would not alter?

Multinationals often invest in impoverished countries, but it is debatable whether countries benefit from such investment. On the negative side, multinationals invest only because labour and material are inexpensive, and they will move elsewhere if costs increase. Multinationals often expect significant government assistance if they locate in a country, yet they have no fixed investment in the country or in its development, and the profits generated by the investment are usually repatriated out of the country. There is little real long-term benefit to the development of the country or its economy; the only benefit is the short-term one of the low wages paid to employees. A more positive argument can also be put forward. Multinationals will source as much as possible from the country in which a production facility is located, simply because it is cheaper, thus they will support the development of local suppliers. Through this, some of the local employees will acquire skills that can be used to generate other employment. Furthermore, multinational investment can be used to help defray the cost of indigenous investment in infrastructure.

There is no clear answer about the benefit of such investment, and, in Ireland, views have changed over time. Initially, the Irish government supported foreign direct investment indiscriminately, assuming that there were would be immediate employment gain and eventual skills transfers which would encourage indigenous economic development. The emphasis changed in the 1990s, to encourage foreign firms that would create high value operations (such as marketing, research and development, training, and other functions) in Ireland, as well as targeting specific industries such as biomedical and software (FitzGerald 2000; Ó Riain and O'Connell 2000). The 2004 budget provisions moved even further along this route by providing tax incentives for research and development activities.

In so far as governments wish to encourage investment, there is much that can be done, apart from providing financial incentives. The computer and telecommunications infrastructure of a country can be an important determinant of investment decisions. A country with cheap labour costs is only cost effective if the production facility can be effectively integrated into the global manufacturing process. If the factory produces manufactured goods, then raw materials have to arrive, but they also have to arrive according to a schedule, and this requires dependable and inexpensive transport, but also dependable and inexpensive telecommunications. The same is true for the output of such a factory. The telecommunications system must be particularly dependable and inexpensive for an 'information' facility since the raw material (computer problems to be solved, information to be input) has to arrive via digital networks and the output (software, information in databases) is delivered via such networks. A production facility where labour is cheap but telecommunications undependable is more expensive in the long run than a facility in which labour is slightly more expensive but telecommunications more dependable and affordable. Similar issues exist with regard to governments investing in training schemes and bearing the cost of providing trained workers for industry in order to encourage investment. Thus countries seeking to encourage investment do so by creating an efficient information infrastructure, complemented by targeted training schemes, and those countries without such an infrastructure will either receive little investment or only investment in which cheap labour is the primary determinant of cost.

A country can also be 'information poor' in developing its own resources. Unable to discover or exploit its own resources, it has to negotiate with multinationals possessing the necessary technology, although the multinationals often negotiate contracts that take profits out of country. Frequently, developing countries have to remain dependent on the middleman to sell their products for the best price. The middlemen make significant profits by matching products with markets. Better information enables the countries themselves to identify possible markets and produce goods for them. This can be especially effective for indigenous cultural products – local carvings, clothing, paintings – that can be very profitable commodities (United Nations Conference on Trade and Development 2002). Access to information can also assist in a country's development by reducing its import costs, for instance, if the country's health service uses new communications technologies to source cheap generic medicines.

Technology cannot alter fundamental global inequalities; poor countries will not, simply by possessing information technology, reach the income levels of the richer countries. But it is clear that the technology can be of assistance, as long as technology investments are sensible. For instance, the cost of providing telephone cables, sufficient for Internet connections for everyone, may be too large a capital investment. However, satellite access to the Internet, from scattered locations, with users coming into the regional centres for access, would be much cheaper. Wireless Internet access in urban areas may be cheaper yet again (e.g. Hammersley 2003). Sustainable development applies to technology as much as any other aspect of development, and inexpensive technologies can often provide significant benefits (see also Bessat 1987; Castells 1998; Katz 1988; Walsham 2001).

The position of poorer people within developing countries can also be improved by new technologies. Traditionally, local producers sell their goods to intermediaries or middlemen who offer them a relatively small price for goods that are then sold for a considerably higher price. Local producers have little choice, as they do not have direct access to buyers, or even the other intermediaries further along the selling chain. However, new technologies can be used to bypass these intermediaries. Individual producers can directly access their customers, or, at the very least, they can directly contact the intermediaries further along the sales chain. This means that the producers can retain a larger proportion of the final price as profit for themselves. Farmers, for instance, can compare the price they are being offered for products with current market values and determine if the offered price is high enough. They can also bypass intermediaries and sell directly to exporters, thus enabling them to retain a higher percentage of the final sale price.[12] The profit for goods valued because of their cultural context can also be greatly enhanced when customers have direct access to the producers. In many cases, even if the technology is too expensive or complicated for producers, they can join together to share their expertise or, if necessary, purchase it. This creates an effective sales and distribution system (see Witchalls 2002). The global information infrastructure can be used to great effect by individuals who are selling 'cultural' commodities such as rugs, carvings and paintings, in addition to those selling agricultural commodities.

8.7 The demise of the nation state?

There are inevitable policy issues to be considered by states, as new technologies reduce the communications costs that previously segregated regions and nations into separate and often isolated geographical areas. All of these policy decisions to be made by governments presume, however, the continued existence of the state, which may itself be in doubt. The very definition of a state has traditionally been expressed in geographical terms – sovereign people in a defined and delimited territory. The boundaries of states have been geographical boundaries, with customs and immigration controls at the border, and with laws that apply within their geographical jurisdiction. New technologies have altered the logic of distance and territory in so many ways. What is the future of the essentially territorial entity that is the state?

Increasing economic and political interdependence reduced the sovereignty of states long before the 'information revolution'. However, the economic globalisation that has followed from information technology has led to national borders becoming permeable: fewer economic, cultural and social policies which affect residents of a state are actually decided by that state. States have entered into bilateral treaties with other states and joined larger political entities; membership in the European Union, the North Atlantic Treaty Organisation, the United Nations, the World Trade Organisation all imply common laws and some diminution of sovereignty of the part of individual states. These entities presume themselves to be composed of sovereign states (since treaties are required for states to function as members of such entities), but such sovereignty is increasingly a facade. Economic activities are increasingly regulated by international bodies, and acceded to by national governments. Member states could choose to reject the regulations of such international entities, but if they did, they would then suffer the consequences, which would often lead to capital leaving the country and investments going elsewhere. Unfavourable tax regimes lead to rapid capital outflows, leaving government economic policies at the mercy of transnational corporations. Labour flows also respond to global market fluctuations, as people move from one country to another in search of employment. The economic cost of exercising sovereignty is often too high, and so states agree to the decisions of these bodies.

Developments in information technology have accelerated co-ordination in global production, distribution and consumption. Most organisations operate in locations throughout the world, shifting factories, personnel and distribution from one location to another. Technology makes it easy for goods, money and labour to virtually cross borders at will. With Internet, cable, telephone and satellite technologies, governments cannot control, much less monitor, information going into or out of the country, which makes censorship impossible. Money, labour and now information flows across borders, as though the borders barely exist.

When states no longer control information, capital, or labour flows within states or between states, what then of sovereignty? Is the state, as the focus for policy, becoming irrelevant? We have already seen that regulation of Internet content is by governments agreeing on common laws. The general move is towards international conventions, on issues such as economic trade, terrorism, child protection, marriage and divorce, consumer protection. Countries agree to implement agreements so that, in essence, the same laws apply in many different countries. This suggests the emergence of a global information structure of policy formation and administration, with common laws and little national sovereignty. New technologies also permit global policy participation as well as global administration; the anti-globalisation demonstrations represent political participation on a global rather than national scale and are possible only by virtue of new technologies. For such people, as well, the state is no longer the centre of policy making; their attention, instead, is focused on transnational and international agencies.

Complementing the development of global systems of participation and administration, new technologies may well encourage the creation of multiple, segmented, diverse spheres, which also challenge the traditional nation state. In these 'spaces', like meets like and difference is not tolerated. If individuals do not like what they are hearing, either from individuals or organisations, they simply ignore them or leave, and find others with whom to interact. This extends to news reporting, where people can chose to listen to the news outlets that share the same values as the people themselves. Rather than being forced to listen to information that challenges or disconcerts, people may only receive information that confirms already held opinions. This has implications for national and international policy, since it implies the demise of the public sphere (Habermas 1989) which has been central to the formulation of public policy. Individuals with differing views previously debated issues and created a political consensus about public policy. This consensus was founded on the necessity for individuals to interact and debate with each other, so appreciating differing views and compromising their own views. The concern is that new technologies will encourage less, rather than more, contact and interaction between people of different backgrounds, thus hastening the demise of consensus politics.

The increasing internal differentiation of the nation state is not a consequence of new communication technologies alone; the flight to suburbs in industrial societies is a means by which individuals already create communities composed of those with similar class and ethnic backgrounds, and individuals' social networks are often restricted along social and economic lines. However, new technologies provide an additional means by which a rather fragile unified culture becomes a series of fragmented and isolated cultures. Satellite and cable television, electronic magazines and journals, live radio and video streams on the Internet and movies 'on demand' all enable individuals in the same locality to participate in starkly different cultures. It may also decrease the amount of individual participation in local face-to-face voluntary activities, thus further undermining a sense of collective identity and trust (see Putnam 2000). This diversified localism will be explored in more detail in chapter 10, but it is important to note, in this context, its contrast with the unifying effect of earlier mass media in supporting the development of national identity (Anderson 1991).

In the face of such fragmentation, it may be the international political and economic elites that provide the input into policy, rather than national elites informed by a local citizenry. Instead of the sovereign nation state, there will be groups of like-minded people whose membership is not territorially delimited, and administrative and political structures equally devoid of any territorial foundation. There may be limited scope for states defined by geographical boundaries to have policy interventions in the aftermath of the digital revolution, even though there are important policy decisions to be made about a multitude of issues such as rural development, distance education, equality in access to information resources, and conditions and location of work and employment, as well as a host of other issues to be discussed in the next chapters.

Chapter 9

Individuals and social change

In recent chapters, the discussion of the digital revolution and information society has moved from technology, economics and politics to broader social issues such as rural development, lifelong learning, and working from home. Such issues are crucial, since to lose sight of the social dimension is to reduce the information society to computers and the market. This is a social transformation – a transformation in the way people live, the way they relate to either other, and they way they perceive the world at large – or else it does not warrant the attention that it has received. How are individuals' lives outside work changing, and are these changes significant or superficial?

9.1 Domestic technology

One transformation, briefly discussed in chapter 8, is the collapse of the domestic/private life versus public life distinction. These were, until recently, distinctly different domains. 'Domestic' space was private and individuals controlled what happened in their home, who could come in, and how activities were organised. Public space was beyond the control of single individuals; it was unpredictable and unregulated. There were clear boundaries separating the two spaces ('my home is my castle'), with different activities in the different spheres. New technologies are now blurring this distinction. Working from home is one example of this blurring, but this is only one of a host of 'displaced' activities. Tele-shopping, tele-banking, watching movies on cable are all activities previously associated with a shared public space but are now possible in an individual private space. As such activities move into the domestic sphere, people incorporate the technologies into the routines of everyday life, and transferring a previously 'public' activity to the home environment alters the activity (see Silverstone and Hirsch 1992 on 'domesticating' technology). It may be possible to watch a movie at home, but not it is not the same experience: there are different reasons for viewing movies in home rather than in a cinema, with different purposes to be achieved. It is possible to shop from home, but, again, the experience is totally different. One does not meet people unexpectedly while shopping at home, one is less likely to browse the shelves and be tempted by new items. The act of travel to a different location is avoided, but so also are all the new impressions or experiences that stepping out of the door may provide (see also Forester 1987; Miles, Bessant et al. 1987).

Technology may be neutral, but its use is social and therefore technology is itself social. All objects have a biography, based on their social context:

> The biography of a car in Africa would reveal a wealth of cultural data: the way it was acquired, how and from whom the money was assembled to pay for it, the relationship of the seller to the buyer, the uses to which the car is regularly put, the identity of its most frequent passengers and of those who borrow it, the frequency of borrowing, the garages to which it is taken and the owner's relation to the mechanics, the movement of the car from hand to hand and over the years, and in the end, when the car collapses, the final disposition of its remains. All of these details would reveal an entirely different biography from that of a middle-class American, or Navajo, or French peasant car' (Kopytoff 1986).

All objects in domestic space demonstrate, as they are bought and used, a web of social relations. Thus, to discuss the impact of technology in the home, one looks at the social patterns. Who is using the technology? What are other people doing at the time? How was it purchased or consumed?

What activities will take place after this domestic information revolution? Commercial organisations hope that convergence, along with high-speed data access, will mean a range of new services that will provide new sales markets: movies on demand, teleshopping, telebanking, and direct access to booking services (people checking a train timetable, booking theatre tickets directly, making plane reservations on line). These benefit commercial organisations because they increase sales and decrease labour costs, thus improving profit margins. However, the use of ICTs in the home is disembedding economic activities from a previous social context, such as shopping without social interaction. What are the benefits for consumers that make such activities popular? For some, the lack of social interaction is itself sufficient benefit. For others, there may be obvious convenience benefits, such as being able to bank when they do not have time to stand in a bank queue.[1] People may not able to shop when the shops are open, or simply do not have the spare time to shop, so it is worth paying to have goods delivered to them.[2] There may be a clear cost benefit if online shopping is cheaper (airline tickets at cut-rate prices are one example). There may be products that have to be delivered immediately, such as online orders of flowers that must arrive the next day. Or, it may be safer to order online than in person; Internet-ordered, home delivered shopping is of great benefit for elderly who are too frail or too afraid to go out. On the whole, it tends to work best for standard products that consumers do not have to see or try on.

This tends to decrease the occasions in which individuals see other people and feel part of a larger community, which is a social price for this new domestication of economic life. That is not to say, though, that new technologies cannot provide a social benefit, just as the telephone did in an earlier century. For those who are unable to go out, chat rooms and discussion lists provide friendship and emotional support, and access to medical websites provide concrete information. People who suffer from an uncommon condition (or do not want to admit to suffering such a condition) may feel isolated in their local community, but can find other people in a similar situation by going online. As mobility becomes a fact of global life, such technologies become a useful means of keeping in touch with friends and family. There are many people whose first experience of computer technology is the use of electronic mail to keep in contact with children and grandchildren.

We take some behavioural changes that have resulted from technology for granted. For instance, the idea of eating dinner while watching television is a change in behaviour; the change is great enough to have led to the development of a new cuisine: the TV dinner. There are many studies regarding changes in domestic life as a result of television (e.g. Silverstone, Hirsch et al. 1991; Silverstone 1994). Instead of segregated activities in different rooms (washing dishes in one room, studying in another, reading the newspaper in another), there is now a more common life, as all sit around the television and organise the rest of their evening activities around the television schedule. A common life does not imply common attitudes, however, as different members of the family have different attitudes towards, and uses of, technology (Silverstone and Hirsch 1992). And now, with the decreasing cost of televisions and increases in family incomes, there are families with multiple televisions in multiple rooms. Instead of children and parents watching the same TV programme on the single TV that offered only a limited choice of channels, there is a great diversity of programmes being watched by different members of the family in different rooms. There is less overlap between parents and children in their worlds of electronic experience, and less control by parents of children's information intake.

Changed technology leads to changed domestic power structures. Previously, one phone meant that members of a family had to negotiate use of the common resource. As a by-product of this negotiation, parents knew who their children were talking to and when. With the development of separate phone lines for teens in some US homes, parents began to lose knowledge of their children's interactions, although they could still monitor usage and phone numbers on the bills. With the advent of teens owning mobile phones and being able to affordably 'text' each other, children can now communicate without parental approval or even parental knowledge. A whole new social network develops which parents can neither monitor nor control. And with the rise of texting each other, just to stay in touch is a manifestation of a new social dependence. Like primates grooming, incessant texting provides reassurance and emotional support, showing the emotional strength of the community that is being created. This is a community that exists in parallel to the adult one, instead of two communities which previously intersected each other as parents and children fought for use of the single phone. The rise of chat rooms and unmonitored computer access from home means social groups are not even restricted to the immediate locality of children's school friends. Children used to have unmonitored contact with peers (e.g. 'hanging out at the mall'[3]), but at least teenagers could then meet only others in the same locality, and only during specific time periods (e.g. the afternoon). With texting and chat rooms, even the restrictions of time and distance on the creation of children's peer communities have been eliminated.

9.2 Family dislocation

We live in a world where movement is an inevitable consequence of a global information economy. Corporations expect individuals to move to where they are needed, and individuals have go where they can find employment suitable for their specialisation (or else settle for a lower salary). Individuals may also have to move to gain new experience and expertise that improves their training and heightens their promotion chances. The human cost is to individuals who have moved away from their extended family, and, as they continue to move, also continue to leave friends behind them. Even people considered to be part of one family may be scattered, as there is an increase in bi-local families (families with spouses working in different locations). We have all become familiar with the use of electronic mail, mobile telephones, and the World Wide Web as a means of individuals keeping in touch in this new scattered world. Electronic mail is not as effective as telephones for providing emotional support, but it is cheaper for very long distances, and also more effective for keeping in touch with people who live in different time zones and who may be home at unpredictable times.

Although these technologies solve the problem of maintaining contacts in a world where people are often quite distant from each other, they have to be understood in their wider industrial context. The technology that produces these communications technologies (including traditional telephones) is the same technology that led to families moving out to suburbs and individuals moving halfway across the globe in the first place. New technologies led to the globalisation of production, distribution and consumption, which in turn led to changed organisational structures, which in turn required the movements of families and individuals. If it had not been for technology that made it possible for an industrial production system to be dispersed throughout the world, there would have been no need for a communication system to keep individuals in contact. This is the continuing story of technology: solving problems that exist only as a by-product of introducing technology in the first place. These communications technologies are merely enabling the recreation of family and social relations that were disrupted by technology-induced economic change.

Although new technologies often capture the imagination, traditional means of maintaining contact remain crucial. The telephone has grown in importance, since reductions in telecommunications charges make cross-continent or even global telephone calls affordable on a weekly – and, for some even on a daily – basis. Improvements in technology have also reduced the cost of physical transport, so that people can afford to travel more often than before. The days of someone emigrating from Ireland, never to be seen or heard from again, are over. Even before the advent of the Internet, for instance, Irish illegals in the United States in the late 1980s kept in touch by meeting people recently over from Ireland, reading regional and national papers, watching sports matches on video or via satellite. This allowed 'the immigrant community to continue to identify instantaneously with national and international events and issues from a distance . . . [and] the "psychological" distance from home is minimised by the range of mass media available to them through the neighbourhood bars' (Corcoran 1993: 104). However, with the advent of new technologies such as electronic newspapers and radio broadcasts on the web, participation no longer requires going to the local bar. At home or at work, people use email and the web to keep in touch with other people and events at a distance. The constantly decreasing cost of telephone calls and travel has increased the amount of communication and further enhances that participation.

Importantly, new communication technologies are location independent. Postal mail and traditional telephone lines connected places rather than people, and people had to be at the expected location to be contacted. If people were not at home, or had moved, contact was lost. New communication technologies are not dependent on location; if someone has changed homes, jobs, or even countries, their email address stays the same, their personal web address stays the same, their mobile phone number stays the same, and (with the introduction of voice-over-Internet telephony) even the traditional land line phone number might soon remain the same. If someone moves locations during the day, over the course of weeks or permanently, he or she can still be contacted. This change from location to person-based communication now means that people can remain in contact over a lifetime, regardless of changes in circumstances.[4] This encourages more intensive and frequent interaction, almost to the point where (especially for teenagers) not receiving messages or phone calls is cause for anxiety.

Person-based communication facilitates the creation of virtual extended families. Cousins scattered throughout the world can now use the World Wide Web as a message board, as a family album for recording life events, and for organising real meetings. As early as 1993, newspapers could report examples such as Fetternet, a family network of 50 relatives run over CompuServe, describing them as a close-knit group which stayed in daily contact through e-mail, despite wide-ranging locations such as Sao Paulo, Brazil; Berkeley, California, Corpus Christi, Texas; and South Bend, Indiana (Garreau 1993). As the penetration of new technologies increases (and the cost of digital photography and streaming video decreases), such virtual family homesteads are likely to increase. In the United States, extended families gather for Thanksgiving; in Ireland, for Christmas. Birth, deaths, and weddings also provide occasions for kin to reassemble in one physical location. These are the ceremonies or rituals that help families maintain themselves. Now, if some of the family cannot attend, a CD with digital photos and digitised commentary may be sent. Or perhaps photos will appear on the family website. Telephone calls, on the day of the event, can help recreate the sense of collective experience.

However, while new technologies often recreate previously existing social networks, there are important differences between the traditional mode of face-to-face communication and electronic communication (including traditional telephone calls). Face-to-face communication is an 'information-rich' mode of communication, whereas the lack of non-verbal cues in electronic mail reduces the effectiveness of communication and increases the chances of misunderstandings (Kiesler, Siegel et al. 1984). Thus, while computer mediated communication may be better than nothing, there are differences in social relations that are conducted electronically. The extent to which social relations can be sustained through electronic rather than face-to-face communication remains to be seen (Wellman, Salaff et al. 1996). Can strong bonds of affection be maintained when contact is largely electronic? Is a virtual Thanksgiving or Christmas as good as the previous assembling of the extended kin? Obviously not (except for those who want to avoid the ceremony in any event). Is it better than no contact or communication at all? Only future research will show how individual and family relations change as communication becomes electronically mediated.

9.3 Individual empowerment

While changed relations with now distant family and friends are an obvious consequence of the demands of the global information economy, there are also less obvious changes in individuals' everyday lives as a result of new work patterns. In the past, everyone went to work at about the same time, came home at about the same time, and relaxed at the same time. Daily lives existed along parallel tracks; it was easy to contact people and it was easy to co-ordinate activities. Work patterns have changed as a result of the global information economy, with people going to work at differing times of the day, for different durations, often in multiple locations, and with little advance planning. No longer is it possible to assume that two people would be off work at the same time and so would be contactable or available for a social gathering. In the face of lives that are no longer synchronised, new technologies enable the same co-ordination in private life that is commonplace in business life. If the person is not at home when someone phones, the answering machine means they can return the call and organise a time when both are available. If the person is unexpectedly out of the home, they can be contacted by mobile phone, regardless of where they are. If the person is not at home when a favourite programme is broadcast, they can record it for later viewing using a VCR. New technologies permit participation and interaction despite a new asynchronous life.

This transition to electronic, rather than face-to-face, communication has led to important changes in social interaction and participation. Interaction with others was formerly unavoidable – one met people on the street, or they called at the house. It was very difficult to avoid meeting someone, whether you wanted to or not, as one shopped, socialised, went to work, or even relaxed. For many, growing up in a village or small town meant a series of these involuntary relationships and interactions. This lack of privacy and lack of control over interaction can become very irritating. Technologically mediated communication (including telephone, as well as email) allows individuals to decide whether to interact or not. They can use caller id to see who is phoning, or use the answering machine to screen calls. If a message has been left, people can return the call at a time of their own choosing (when they feel rested or able for interaction). For many, this is far better than having to talk whether they like it or not. One can even pretend there is a malfunction in the answering machine, and not return the call at all. Similarly, electronic mail messages can be read at one's own leisure, and either ignored, or responded to at a later time. If desired, one can shop or bank from home, and avoid interactions with people on the street. Social interaction is by choice not necessity, and there has been a move from involuntary social relations to voluntary ones. People can avoid contact and so communicate only with those they wish to. This decreases people's ability to deal with conflict. Two recent studies of Internet use confirm that people intentionally use electronic mail to avoid saying unpleasant things to friends or relations (PEW Internet and American Life Project 2000; 2001a). Instead of confronting conflict and disagreement face-to-face, or even over the telephone, people preferred technologies that meant that unpleasant responses could be ignored. There is a real danger that individuals could forget how to deal with interpersonal situations involving conflict or disagreement.

Such technology can also be used to communicate with people previously inaccessible and, potentially, create new relationships. We are all now familiar with electronic chat rooms, in which strangers can strike up 'conversations' with others. While chat rooms have become associated for many people with unsavoury activities (such as adults using chat rooms to exploit children), this leisure activity can also become the basis for long-term beneficial social relationships. People who feel isolated in their locality can make electronic contacts that provide an important remedy to their isolation. Imagine someone suffering from a disease. In some cases, it might be an illness that they would not want to reveal to others, whether cancer or AIDs or a learning disability. Perhaps they do not want their neighbours to know about it, but still need someone to talk with. Suppose, even if they did want to talk with someone, there was no one in the locality who know what the disease was or could understand the impact that it was having? Such people can now find others, anywhere in the world, who share the disease and understand its impact. These are people with whom one can share experiences, ask questions, complain to, and get emotional support from. Anyone who feels isolated and alone can electronically find emotional support, as well as information, that they could not obtain in a face-to-face community. Individuals who felt different and alone no longer have to suffer that isolation; they can break free and create new relationships in a wider world.

New technology has been enabling individuals to exercise more control, choice and autonomy; it can allow greater interaction and communication or isolation and privacy. People usually combine both, sometimes using mobile phones and answering machines to maintain contact, while at other times using the same technologies to avoid contact. This is part of a general trend towards lifestyle by choice rather than accident. Assuming sufficient access to economic resources (money), one can purchase the freedom to choose the personal identity that one wishes to present to the world. This is obvious when communication is via the computer, where people can choose false nicknames, but it is even more true in everyday interactions. Technologies such as contact lenses, liposuction, tanning beds, hair colouring, are all used to create the identity presented in face-to-face communication. With increased buying power and decreasing cost of consumer goods, additional identity choices are available through the clothes one wears, the food one eats, the recreational and entertainment choices one makes. New technologies make such options cheaper to exercise and more effective.[5]

Research in the United States gives a vivid picture of the extent to which people live their lives using new technology. As a recent survey of Internet use in the United States suggests, the Internet is becoming an integral part of social life. The following percentages of Internet users said that the Internet played a crucial/important role in making the following life decisions:

- 29 per cent who obtained additional education or training for their career
- 27 per cent who bought a car
- 25 per cent who changed jobs
- 22 per cent who found a new place to live

With regard to personal lives, the following percentages of Internet users said that the Internet played a crucial/important role:

- 24 per cent of those who dealt with a major illness themselves
- 26 per cent of those who helped another person deal with a major illness
- 33 per cent of those who started a new hobby say their use of the Internet
- 15 per cent of those who started a new romantic relationship
- 15 per cent of those who ended a romantic relationship
- 14 per cent of those who got married

Those statistics (PEW Internet and American Life Project 2002) portray a picture of everyday life in which new technologies are as important as the telephone or face-to-face communication. The same research group reports that new technologies are an integral part of the communications resources used to keep in touch with family (PEW Internet and American Life Project 2000), with over half of Internet users reporting they have more contact with their family than before. Indeed, research from the same organisation (PEW Internet and American Life Project 2001a) of teenage use of new technologies in the United States paints a picture of a generation that shifts from one mode of communication to another constantly and effortlessly.

9.4 Autonomy or subservience?

Although new technologies allow increased control and autonomy, there is a disguised price. A more accurate description might be the illusion of freedom, autonomy and choice, but the reality of dependence and control, with the increased choice only about superficial things. Consumer benefits, such as computers, mobile phones, fashionable clothing, exotic foods, are only available to those who earn sufficient income from participation in a world economic system. We have little choice but to work, and are often able to make very few decisions about such fundamental issues as the kind of work available, and the salary or location of it. We are all enmeshed in a market economy that we are unable to opt out of. It is possible that there is decreasing choice about fundamental issues over which individuals have little control and little policy input.

Our dependence is technological as well as economic. Traffic lights, dishwashers, answering machines, mobile phones, cars – all of these are technology activities in which embedded chips with pre-programmed software limit and constrain choice. How often, when using a word processor, do we change our behaviour to suit the programme? Yet opting out of such control is difficult, if not impossible: one cannot ignore traffic lights, one cannot re-programme the dishwasher, and one cannot avoid word processors. One can hardly remain in office employment if unable or unwilling to use electronic mail (and promotion may be linked to technological proficiency). People who do not have a mobile phone may find themselves without a social life, since social life is more impromptu than formerly and it is difficult to organise events without mobile phones. With embedded chips everywhere, more of our behaviour is governed by technology, and we are ceding control to those who design the technology.

Furthermore, many of these technologies have an element of surveillance built into them. Mobile phones enable companies to track movements of individuals with increasing accuracy. Credit card transactions allow the tracking of purchases. Supermarket loyalty cards enable organisations to link individuals with specific commodity preferences. There may be an increased choice in information content and people with whom one communicates, but this requires standardisation of hardware and software sold by international providers such as Microsoft, Dell, Compaq/ Hewlett Packard or IBM (who are themselves usually based in the United States). Increasingly, our lives are governed by technology that reduces our freedom and privacy, and which we are unable to opt out of.

As economic activities become disembedded from social life, it is easier to accomplish more things during a day without human interaction. One can shop for goods, pay bills, watch a movie, book a train ticket, all without dealing with human beings. As social skills decrease and stress levels rise, machine interaction becomes easier because it is more predictable and more controllable. There is a danger that individuals will thus develop a preference for interacting with machines rather than people. Silverstone (1994) has already suggested that children can find the television more reassuring than individuals: it is always available, it does not get moody, and it does not respond negatively to the child's behaviour. Similar concerns have been raised about children playing video games (Greenfield 1984; Turkle 1995; Turkle 1984). Many of our human skills focus on dealing with conflict and learning to live with diversity; we learned to get on with people we disliked because we had to work with them or we had to live near them. If we lose the social skills that develop out of such situations of unavoidable interaction, will we begin to prefer machines over humans?

There has also been a subtle change in the nature and range of social relations as a result of new technologies. People have close friends or kin with whom they keep in touch, regardless of changes in employment, residence or marital status. Then there are acquaintances that are easier to lose track of. Such contacts constitute 'weak ties' (Granovetter 1973; 1982); they are people one might go to for information, but would not ask for significant assistance. In the 'wired' society of global communication, individuals have links to other individuals scattered all over the world, following, almost physically, the spaghetti-like growth of wires connecting computers worldwide. We keep in contact with distant people and places, when we would previously have lost contact with both. We transfer messages amongst friends, so that friends of friends themselves become friends, and we have an impressive range of latent acquaintances that can be called on, as necessary (Wellman, Salaff et al. 1996). There has even been a growth in websites that will record an individual's name, current email, and schools attended, so that people who attended the same school at the same times can contact each other. Educational institutions, especially universities, are themselves beginning to encourage former students to maintain contact with each other, especially through the institution; such groups can be useful benefactors to the institution. There has been an impressive growth in the number of 'weak ties' which people can maintain using new technologies (and often using electronic diaries to maintain the list of such people).

These changes in participation and communication are not restricted to small elites; evidence indicates that an increasing percentage of the population of most industrial societies participates in these new technologies. Seventy-five per cent of Irish households possess a mobile phone, residential Internet access is nearly 36 per cent, home computer ownership is 42 per cent, and 33 per cent of households own a DVD player (Central Statistics Office 2003a). These rates are not the highest in the world (Norway, Sweden, Italy, Portugal, Iceland and Luxembourg all have mobile phone rates above 84 per cent), while, in terms of Internet usage, Austria (48 per cent) Finland (53 per cent), Luxembourg (54 per cent), and Sweden, Denmark, The Netherlands (about 64 per cent) are all higher than Ireland. However, Irish usage rates are still significant, considering that Internet usage in the United States is 60 per cent, and the Irish mobile phone rate is significantly above the US rate of 60 per cent (http://www.nua.ie). Most significant is the rate of increase. Mobile phone usage in Ireland rose from 43 per cent in the first quarter of 2000 to 78 per cent in third quarter of 2002. Furthermore, as of February 2001, 95 per cent of 15–24 year olds had the use of a mobile phone. Use of the short message service (SMS or 'texting') rose from under one million messages during the first three months of 2000, to just under six million in the latter three months of 2002.[6] Internet usage rates have nearly doubled in the same period of time (http://www.nua.ie). This is a transformation that is not restricted to a small segment of the industrial world and the social consequences will apply across gender and class divisions.[7]

9.5 Virtual reality

Technology has blurred the division between home and work, it has removed the limitations of place from interaction, and it has even blurred the very distinction between reality and illusion. Everyday experience has always had an authenticity that could not be questioned; it combined visual, audio, and tactile information that could not be faked and so was authoritative because it was authentic. It is the ultimate multi-media environment, whereas technologically mediated communication is restricted to single media with limited information content (radio plays audio information only; television relies largely on visual images). Electronic information could not masquerade as anything other than a representation of reality, whereas 'authentic' reality was transparently obvious. Virtual reality technologies which try to blur this by creating convincing simulations of real experience have always been derided: kids playing on video machines, pilots in flight simulation cabins, surgeons training on virtual bodies engaged in simulation that are transparently not real. Only desperate or gullible people would accept this substitute for everyday life as 'real'. Thus, there is the technological world of imagination, and the everyday world of experience, and never the twain shall meet.

In fact, we all live in virtual realities. The term 'virtual' refers to the essence of an object or action. A person is the 'virtual' ruler or decision maker because he or she is the actual or effective ruler, whatever the formal trappings, or outward appearance. If something is 'virtually' true (such as the company closing or the employee being laid off), then it is true in essence, even if not so in appearance. So, virtual reality abstracts the essential bits of reality and re-presents them. In a way, maps and scale models are a 'virtual' reality – they are a representation that has captured the essential characteristics. A virtual reality is never expected to be a complete representation of experience, just a representation of the essential or crucial bits of experience.

Part of the process of experiencing such 'essential' aspects of reality is to suspend disbelief, often aided by technology. Voluntarily entering into an artificial experience is nothing new. This ability and willingness to suspend disbelief and enter into an imaginary world has been part of human life for centuries, with stories at bedtime, books, movies and plays. People have not only always suspended their disbelief but have assisted in making the process easier. People watch movies in a specially darkened room with a large screen and high quality speakers because it is easier to 'lose oneself' in the theatre.

People read books but turn off the radio, so as to lose themselves in the book more easily. The significance of new technologies has been to make it easier to suspend disbelief because the electronic creation is becoming more convincing. New technologies are increasingly multi-media and authentic, and the boundary between reality and illusion is becoming less clear-cut, especially as many new technologies start with 'authentic' images that are then manipulated.

This has a number of implications. We are already dependent on technology for our picture of the outside world: we look at visual footage on television and believe that we are seeing real events. At one point, such images had to be real because we lacked the technology to create convincing replicas. However, the digital technologies that now present reality (e.g. photos and TV coverage) lead to those representations being manipulated. Real photos can be altered, false photos can be made to look real, and entire movies can look real, but actually be made with computer animation. Gone are the days of watching a science fiction movie and seeing the zip of the costume, or watching a news event and seeing the participants being prompted to engage in theatrically fake actions. It is perfectly possible for people to look at images on a screen which are deemed 'real' because they are part of a news programme, even if the images are fabricated. People make the leap of faith and imagination to see them as real and react to them as 'real' events. They trust that journalists and news editors have vetted the images and deemed them authentic. We are increasingly dependent on the context of information as a means of authenticating it: it is real not because of any internal authenticity, but because it is validated by its context of being part of a news programme. Yet this can go wrong. The fright of the War of the Worlds broadcast by Orson Welles in 1938 was because people believed in the authenticity of radio. It was understood to be a 'live radio broadcast' and so listeners imagined the world described by the broadcast to be real (Koch 1970). One has to trust that the source of information is providing accurate images and sounds; there is no way to tell from the images themselves whether the material is accurate.

It is almost as worrying that people can watch real events and take them to be fake. In the first US-led war against Iraq (Galvin 1994), there was excellent coverage of tracers of missiles over the sky. For many, there was little difference between those images and images in a video game. The visual images that came from the 11 September 2001 events in New York were horrific, but many people initially assumed that they were watching the special effects of a very good movie. Our reality is increasingly one in which telling the difference between reality and illusion is problematic. We can no longer depend on the intrinsic characteristics of the information to decide if it is true or not – real events can look false and false events look real (see Slouka 1995). We trust broadcasters to have verified the material. Increasingly, however, as 'news' broadcasts come from a variety of sources, it is difficult to trust that such material has actually been vetted, and we are also less willing to trust in the accuracy and impartiality of journalists or editors.

The blurring of 'fact' and 'fantasy' is especially dangerous for young people. Adults have learned to usually be able to tell the difference between real and illusory experiences. Although we can be misled if the material is presented in an authoritative manner, most of us can usually guess whether an event 'rings true' or not. When we go to a movie or a play, we know that we are entering a world of illusion; we do not confuse actors with their persona on the stage. However, children playing video games may not be so experienced in telling the difference between reality and illusion. Furthermore, the video games are so convincing in their reality that it may be more difficult for children to learn the difference. All too often, now, children who see disasters on television may not be able to understand that the people being filmed are not going simply to get up afterwards and go back to their everyday lives. This is all the more worrying in the case of children accessing pornographic images or movies on the Internet; the activities may be faked and, whether faked or not, presented as though they were a 'normal' part of everyday life. Children who are still learning what is 'normal' are in no position to dispute such implicit claims when watching pornography. The violence or sexual activity that children see on video may seem transferable to everyday life, as children act out what they have observed electronically. The debate is ongoing about the long-term impact of video games (Greenfield 1984; Herz 1997; Provenzo 1991; Sheff 1994; Turkle 1995; Turkle 1984), but it is obviously a concern for the future.

It is also of concern that adults might find virtual reality to be an all too satisfactory alternative to the problems of everyday life. Just as electronic mail is not a convincing alternative to face-to-face communication but sometimes better than nothing, virtual reality may be unconvincing but still better than the alternatives. There tends to be an assumption that technologically mediated communication could never be as 'real' or satisfactory as face-to-face communication, but the truth is more complex. People find themselves emotionally involved in, and moved by, books they read or movies they watch. It is possible to have a significant emotional reaction to a fabricated and artificial experience, depending on the extent to which one 'suspends disbelief'. People are able to project emotional values onto the relatively 'flat' communication content of a letter or phone call (as compared with the richer content of face-to-face communication). Family members are well able to sustain emotional relations via telephone or letter, despite the technological limitations of each. Technology is making it easier to recreate the media richness of face-to-face communication, thus making it easier to project emotional or personal content onto messages.

New technology also enables people to presume familiarity and intimacy with strangers. Public figures sometimes encourage this, as when Franklin Roosevelt, President of the United States, had 'fireside chats' with the nation over the radio in the 1930s (Beniger 1986). This enabled people to feel they had a personal relationship with a distant and impersonal government. Many media personalities cultivate this sense of 'personal presence' to sell their products and thus enhance their commercial value; they may encourage stories of their personal life to appear in public press; they may hold 'chats' on the Internet, or they may encourage personal websites. Sometimes there is an illusion of a two-way communication flow, with some email messages being answered or some personal appearances being made. Individuals themselves may want to claim intimacy and familiarity, as though they personally 'know' news presenters or soap opera stars. Such a sense of personal contact may actually be a good thing, especially in so far as it gives people a sense of personal participation in an increasingly distant political and social system.

The debate on technologically mediated communication and personal relationships continues. Some suggest that individuals will use technologically mediated communication to create and maintain emotional relations that are 'virtually' [pun intended] the same as ones that depend on face-to-face communication. Others argue that the limitations of technology will always constitute a barrier to such rich relationships. It is not yet clear whether these technologies will make it easier for people to maintain rich emotional relations with distant family and friends, or simply serve as a complement to existing face-to-face relations. The anecdotal evidence, however, suggests that some people are quite able to project strong emotional feelings onto the messages that circulate on the Internet. Stories abound of long lost friends reuniting over the Internet, and even long lost loves that re-establish their relationship via email and mobile phone calls (Weale 2002).

It is clear that technologically mediated communication differs from face-to-face communication not only in its content, but, as with any technology, how it is appropriated into everyday life. For example, it permits private communication without observation. It is difficult to 'meet' someone in a pub or party without spouse, friends, or family knowing, but it is relatively simple to meet the same someone privately and electronically. It may also take quite a long time to exchange personal information, when face-to-face meetings are short and unpredictable; quite a lot of information can be exchanged during solitary and uninterrupted sojourns in front of a computer. Verbal speech did not replace non-verbal communication and cannot be analysed in the same way as non-verbal communication and writing is similarly different from speech; technologically mediated communication is not a replacement for face-to-face communication, but a hybrid that must be analysed in its own terms (Lea 1992; Mantovani 1996).

9.6 Reality and authority

The discussion of virtual reality emphasised our dependence on authorities to help us decide if news, or any other information, is 'authentic' or not. This is true of information on the World Wide Web, where fake or unsupported claims can look as accurate as well documented claims; we are dependent on the source of the information to determine the accuracy of the information. This dependence on authoritative sources is developing just as the very basis of 'authoritative' sources is being questioned. The Information Age has witnessed an increased amount of information in circulation, but also an increased diversity of information. With increased choice, there is now conflicting information circulating on any issue or any perspective under the sun. When formerly faced with conflicting information, people trusted science as a single authoritative source. However, recent decades have seen a decline in people's trust. Scientists were once seen as disinterested researchers, and science as proclaiming undisputed facts. Now, scientists are often seen as having their own vested interests, and science as responsible for getting things wrong as often as right. For some, science increases the extent to which we are all at risk. Pesticides which were supposed to lead to greater food production at less cost are now seen as poisoning both the environment in general and food in particular. Nuclear energy was supposed to be a safe and clean source of energy, but is now seen as creating residues that will take centuries to disappear. Opposing factions now find 'scientific evidence' as well as scientists to support their position; in the 'right to life'/abortion referendum in Ireland in March 2002, opposing sides were able to find both scientists and scientific facts to bolster their opinion. Since both parties claimed the authority of science in media reports (television, radio, newspaper reports), voters had to evaluate these claims and counter-claims using other criteria.

The perception of danger and risk has been increasing while the credibility of scientific authorities decreases.[8] Medical information illustrates the demise of a uniform, authoritative, viewpoint. In the past, one went to a local doctor who made an accurate and impartial diagnosis and then provided an up-to-date and relevant treatment. Now, some people see doctors as prescribing medication simply to satisfy the patient or because the doctor has been encouraged by drug companies to use a particular medication. Some doctors may make accurate diagnoses, but others may make mistakes; some may provide the best current treatments, but others may prescribe inappropriate or outmoded treatments. People often use the Internet to reduce their dependence on local doctors, and try to make their own diagnosis or look for alternative treatments. The local doctor is no longer the single source of information about all medical matters, and it is the patient who decides which source of information is to be used. While some doctors encourage patients to seek alternatives, and then participate with the patient in determining the best course of action, it is often the patient who makes the decision. Sometimes the decision is accurate, but the patient just as likely to diagnose the condition inaccurately or decide on an inappropriate treatment, with possibly serious consequences for the patient's health.

We live in a differentiated society in which we are all dependent on specialist information outside our own control and knowledge. Yet people are less trusting of others as providers of accurate information. They decide for themselves what constitutes an authority, often with insufficient knowledge or experience of the area on which to base a decision. We all have the experience of asking friends for a 'good' plumber or a 'good' doctor, or trying to lever one's own position or connections to get 'better' treatment; these are consequences of combining dependence with insufficient experience to evaluate experts. There has been an increase in the number of websites that are intended to provide users with exactly such expert opinion: medical sites to help diagnose illness, computer sites to help buy the best computer, financial sites to help make investment decisions. There are even sites which record and collate opinions about service providers (plumbers, electricians) in the local community, so the individual can find a 'good' plumber if there are no recommendations from friends or family (Power 2001). New technology is trying to overcome yet another problem that is a by-product of the digital revolution: how can individuals find and evaluate the information on which they are increasingly dependent when living in a diverse and conflicting information universe?

In this chapter, changes in the ways individuals relate and interact in the home have been explored, as have changes in the ways individuals participate in society. Individuals are now more dependent on information, but the digital revolution has also changed the way individuals evaluate information; there is decreased trust in authoritative sources but few alternative strategies available. With the increased diversity of information and experience afforded by new technologies, what of the collective experiences of community, identity and culture? The next chapter will explore these dimensions of the information revolution.

Chapter 10

Beyond the individual: culture, nationalism, community

Economic and social changes over the past fifty years seem to have forced people to live increasingly isolated lives. Since the Second World War, people have left their urban neighbourhoods to live in anonymous suburbs. They have moved locations as their employers expect them to work anywhere in the world. With new technologies, people work from home, shop from home, bank from home, and even socialise from home. The common experiences that created bonds amongst people in the same place seem to be diminishing. Yet human beings are social creatures; we want to live in webs of social interaction, even if we have to create them ourselves.

10.1 Imagined community and global culture

As social beings, we live in groups. Until recently, these groups were geographically delimited: physical proximity determined our primary social relations, and economic, political and social activities were constrained by spatial limitations. With the digital revolution, location is no longer a constraint and distance no longer an obstacle. Phrases such as 'networked society' and 'global village', often used when discussing the information society, imply linkages that make distant places local. Other images such as cyberspace, electronic town hall and virtual community seek to apply the terms of localism to dispersed networks of electronic relationships. All of these emphasise that physical distance, previously a defining feature of social relations, is now less significant. A common vision of the information society is the 'death of distance' (e.g. Cairncross 1997). What impact will this have on the communities that have been the focus of social life?

Children grow up in a locality, and learn the same history and literature, regardless of which school they are in. They meet each other when playing sports and when 'hanging out', they watch similar television programmes and read similar magazines, they gossip about the same popular figures (whether music, movie or soap opera). Once adult, they are concerned about the state of the local road, or the opening hours of the shop. Interaction creates and maintains a common sense of experience and common identification that we call community.[1]

The creation of a common national identity is more complex. People who believe themselves to be Irish, British or any other nationality share common identification with others, even though they will never have face-to-face contact with most of those whom they see as fellow Irish, British or French. They are part of an 'imagined community' (Anderson 1991) – sharing common values with other people like themselves as though they were part of the same community. Creating and maintaining such an identity is a complex process, but an integral part of the process is the shared experience of similar education, similar mass media content, similar language and public holidays (Hobsbawm and Ranger 1983; Fox 1990; Gellner 1983). Weber (1977) recounts the French experience in the late nineteenth century and notes the impact of the educational system:

> The symbolism of images learned at school created a whole new language and provided common points of reference that straddled regional boundaries . . . [A]ll children became familiar with references or identities that could thereafter be used by the authorities, the press, and the politicians to appeal to them as a single body. Lessons emphasising certain associations bound generations together (p. 337).[2]

Effective as this was for metropolitan France, it is a bit incongruous when Moroccan children would recite, as part of their history lesson, 'Our ancestors the Gauls . . .'. Similarly, Irish people learn a history and language, they celebrate similar holidays and they have similar experiences of politics and government. Furthermore, these are experiences not shared by people who are exposed to different influences by living in other countries. Shared experiences of language and literature, schooling and government structures, all foster the sense of common identity and experience that is 'national identity'. The creation and recreation of national identity has been aided by technologies such as mass newspapers, television, radio and magazines.[3]

If the geographical limitations of mass media technologies have fostered a geographically delimited collective experience and thus identity, then what impact will new technologies have on this process? If new technologies allow information to circulate without geographical limitations, does that mean a common global experience, and thus a global culture? Increasingly, everyone in the world has access to the same information and same commodities: similar movies, television programmes, world-wide news programmes, fashions, children's toys, fast food outlets. If culture is the result of common experience and shared information, then is global culture the inevitable consequence of people participating in the same common experiences? This global culture is a distinctly different proposition than a global information economy, which is the participation of people in a unified economic system, with international structures for the production, distribution and consumption of goods and services. Global culture is the sharing of beliefs and ideas, as well as commodities or artefacts.

There has been a long debate about globalisation (for a flavour of the discussion, see Harvey 1989; Featherstone 1990; Robertson 1992; Featherstone 1995; Featherstone, Lash et al. 1995; Friedman 1994; Appadurai 1996). The development of global communications and its domination by multinational companies led to concern about a monolithic imposition of uniformity. Films produced in the United States are shown everywhere, to the detriment of locally produced films. The US media products are not necessarily better, but simply cheaper to buy. This is partly owing to the economies of scale that arise from producing commodities that can be sold in markets throughout the world. In addition, these products are sometimes 'dumped' on foreign markets. Films have not been the only commodities that circulate in the global economy. Many commodities have been internationalised, especially those relevant to 'fads' and 'styles'. Whether it is fast food, jeans, skateboarding, Power Rangers or Teletubbies, people increasingly have the same buying preferences in clothes, food and music, leading to uniformity in the films and news viewed, the clothes worn, the food eaten, and the music listened to. A concern has arisen that there is a global culture emerging with a monolithic set of meanings, and that since the products are those of the United States, the American cultural values implicit in them will become global cultural values. That is, all cultures will be subsumed within single set of meanings (US meanings) at the expense of local, indigenous cultures (Tomlinson 1991).

This is a real concern for many countries as well as individuals, some of which have instituted regimes to restrict foreign media and encourage indigenous media (see Collins 1996; Preston 2001; and the discussion of the 1995 Green Paper on Broadcasting in Ireland in Kiberd 1997). Especially in countries where broadcasting is controlled or supported by governments, there is a often a restriction on the amount of foreign programmes that can be broadcast, and financial support for the production of local programmes. This is partly to protect or encourage local employment, but also reflects a desire to protect local culture from foreign, usually US, media. In Ireland, for instance, local media production is explicitly linked with protecting and fostering 'Irish' culture, which is seen as being under threat. This attempt at control can sometimes also extend to other elements of global exchange. For instance, foreign music, clothes or food can be highly taxed or otherwise restricted, in order to favour consumption of local items.[4]

This conflict between foreign and indigenous elements is overstated, and dependent on a somewhat simplistic view of the impact of mass media on local societies (Boyd-Barrett, Braham et al. 1987). The interaction between local and global creates a melange or mix of different elements, some internal but some external. Economic factors may distort this mix. For instance, United States viewers are exposed to few non-US media products,[5] while European viewers are more likely to be surrounded by US media products than by the products of their European neighbours. Small wonder that in Ireland there is uncertainty whether the images used to conceptualise Irish society should be drawn from Boston or Berlin. The impact of this disparity is even greater outside Europe. However, regardless of the disproportionate influence of external elements in some societies, these elements are changed as they are incorporated into a local framework. The local meanings associated with these elements may have very little in common with any global meanings, as Wilk (2002) demonstrates in a discussion of the impact of United States generated sports programmes, delivered via satellite, on a Latin American society. Television viewers are not passive recipients of global meanings; the meanings of global cultural products are altered by local circumstances, even within the same society (Morley 1980; 1992; Ang 1996; Silverstone 1994). The television show *Dallas* is not interpreted in Delhi, Dublin and Dallas in the same way; there is an interaction of the cultural product with local values and beliefs (Ang 1985; Liebes and Katz 1990). This is not to say that external products have no impact, but that the elements combine to create a unique mix.

Instead of a uniform global information system, there are varying local versions or interpretations of global material. Similar trends in global news organisations, both print and satellite, have led to a 'localised' version of news for Europe, Asia, South America and other areas. Satellite broadcasts introduce local advertisements for different national markets, even within the same broadcasting 'footprint'. This has also extended to global Internet sites; yahoo.com has been replaced by yahoo.ie, yahoo.fr, and other national sites, each with its own local identity. These are responses to the preferences of users, and the result has been to create local diversity rather than global uniformity.

The interaction between local and global is not restricted to mass media content or information services, but applies to the range of global commodities. One example would be food. On one level, there are national cuisines, such as Chinese food, that are distributed throughout the world. However, 'Chinese food' manifests itself differently in each location. First, the ingredients and seasonings vary, depending on local tastes. More importantly, the food is eaten by different types of people, at different times, on different occasions, and therefore has a quite different social significance in different societies. In some cases it is associated with twenty-somethings, emerging from pubs late at night; in other cases, it is associated with forty-somethings concerned with organic food. Chinese food can become identified with specific activity or specific social or ethnic group. It may be the kind of thing you might do in a group going out for a cheap night of socialising in the UK, while being a quite different event in India! Similar observations could be made about that archetypical symbol of US cultural domination: McDonald's. The social context of French people going to a McDonald's in Paris is different from McDonald's in Dublin or Manchester – the food is slightly different, and even the architectural design is different.

The outcome of local and global interactions in Ireland is complex. Ireland is a producer, as well as a consumer, of global commodities. Ireland has exported the commodity of Irish culture via music (Chieftains, Clancy Brothers), dance (the Riverdance ensemble), clothes (Aran jumpers), pottery and glassware (Waterford Crystal) and obviously literature.[6] Ireland also incorporates products and ideas that emerge from elsewhere. Country and western music has transplanted itself,[7] as has the celebration of Halloween. From France and Italy have come cappuccinos and café society. Dress styles from diverse locations such as the United States and the Middle East are combined together. Ireland becomes a unique local articulation of multiple global flows (Hannerz 1996; 1992). In the circulation of commodities, societies are clearly developing their own cultural specialities, branded in order to maximise global market share.

Local cultures are not under threat from external media alone; there is also the question of internal diversity. Members of a nation have never shared the same experiences and views perfectly; regardless of the rhetoric of national identity and imagined community, there have always been class, regional and ethnic divisions and conflicts. Common national forms and values accentuate the shared elements and minimise this diversity, but the diversity is never eradicated. The domination of national media and national commodities simply restricts the amount of choice that individuals can exercise, which restricts the manifestation of diversity. New technologies have increased the amount of choice and the ease with which action can follow from choice, which has made it easier for existing differences to be expressed. Each person chooses different bits to incorporate into their own unique mélange, which distinguishes him or her from others in the locality. Diversity seems to be increasing, but, in fact, it is a diversity that was previously latent and is now overt.

10.2 National identity and place

If new technologies are undermining a uniform national identity, the outcome of increased cultural flows and increased individual choice is not necessarily one of societies composed of isolated and differentiated individuals. An alternative is individuals participating in multiple collective groups, with no single group having a primary claim on individual identity and loyalty. The fragmentation of mass media (narrowcasting with satellite and cable), combined with increased access to a variety of information sources

(WWW, email, video, etc.) leads to a diminished sense of collective national identity and an easier assertion of alternative collective identities. For instance, in 1977, more than one half of the population of the United Kingdom (28 million people) all did the same thing on Christmas Day: they watched *The Morecambe and Wise Show* (Clarke 2002). This was a collective national experience that is unlikely to be matched in the contemporary world of multiple television households, with satellite and cable choice of programmes. In 2000, the BBC, having purchased the rights to broadcast a major film on Christmas Day, secured only one third as many viewers as in 1977. The alternative choices provided by new technologies mean the end of such common experiences, which were important for their own sake, but also as the basis for subsequent conversations and discussions. In Ireland in the 1980s, the *Late Late Show* was not only a shared national experience of everyone watching the same television show on a Saturday night, it was also the basis for radio and newspaper coverage the next day and personal conversations amongst friends and neighbours for the following week.

What happens when people grow up in a country, acquire a sense of national identity, and then leave? Irish people, for instance, migrating to Britain, Canada, or the United States for work, Jamaicans coming to London, or Turkish people coming to Germany? In the past, people had little scope for maintaining any contact with their home because of high communication and transportation costs. The high cost of telephone calls, as well as the high cost of physical travel back to the home culture meant that the amount of contact, through newspapers or letters, would be minimal. It was difficult to keep in contact with events 'at home', regardless of the strength of the emotional attachments to people or place. There was an inevitable assimilation process, during which the home identity remained more important than the host environment, but everyday contact was either with people in the host culture, or with others who shared a similar expatriate status. The home culture became more remote and the experiences of the host culture led to some degree of acculturation. Thus, we are used to the emergence of 'hybrid' identities, such as Irish-Americans, Pakistani-British, Turkish-German, Algerian-French. Eventually, there would either be assimilation into the host culture, or a sub-culture, such as 'Italian-Americans', would emerge as a stable group that recreated itself generation after generation. These would be people who would maintain many elements of their home culture, but, without dynamic contact with people or events of that culture, these cultural elements would be frozen in time. There are many ethnographic studies of the enclave communities that have developed in such host societies as the United States, Britain and Germany.

With new communications technologies, it is possible to maintain contact with both people and events of one's home culture, even after leaving home.

This was possible to some extent even prior to the digital revolution (as noted in the earlier discussion of illegal Irish-Americans in Corcoran 1993), but is much easier since the mid-1990s. Communication costs are reduced, so that one can maintain contact with friends and family via telephone or electronic mail. The cost of high-speed Internet access and the capital costs of computer ownership have both lessened. Reduced transport costs also means access to 'home' cultural products such as clothes, food, books and movies. There has been a substantial increase in the amount and range of digital media available for worldwide consumption. With electronic newspapers, cable and satellite television, and radio and television stations broadcasting on the web, it is easy to keep in touch with cultural and political events at home. With reduced transport costs, one can afford to travel back to one's home culture relatively easily. It is now possible to participate in the public life of one's home culture, as well as maintain social contacts with friends and relations, despite increased labour mobility (Gillespie 1989 provides an example from London). It is even possible to create cultural products that unify dispersed members of ethnic groups who no longer have a home providing cultural 'source material' (such as displaced tribal peoples exiled from Laos, Schein 2002).

Ireland provides the obvious example of this transformation. In earlier times, people may have wanted to maintain contact and return eventually, but it rarely happened. Air travel was expensive, as were telephone calls, so people drifted away. This inevitable outcome was realised in advance, though rarely admitted, and it led to the 'American wake': a leave-taking ceremony that signified the permanent loss to the Irish community of that person. While such emigrants may not have assimilated completely into American culture, they and their descendants felt out of step if they ever did return. The height of their contact was money sent back to Ireland, either to relations or for 'good causes' (such as the Irish independence movement). This contrasts vividly with the current experiences of an Irish person in the United States. There are electronic Irish newspapers and many Irish radio stations broadcast on the Web. It is increasingly common for radio programmes to report email comments from people all over the world who are listening to the programme live via an Internet broadcast. This is 'real time' national participation and interaction on a global scale. An increasing amount of video output is also available on the Web. Nor is this is simply a national phenomenon; local communities in Ireland increasingly publicise local events on their websites and, amongst the numerous discussion lists and bulletin boards devoted to Irish topics, there are specialised sites for local events. Air travel has become significantly cheaper, and frequent return visits are possible, not only for special events such as weddings, but for the important cultural events such as Christmas, as well as sporting events. The population of Ireland swells on such occasions and these events become important occasions for maintaining social cohesion (as well as personal contacts); since so many other Irish emigrants are home at the same time, these events provide opportunities to maintain contact with other emigrants.

What is the impact of this increased and richer communication? Does more contact with one's home culture and the friends and relations one grew up with just make the separation more difficult to tolerate and the homesickness worse? Does it permit individuals to maintain their participation and identification with their home culture and ignore their immediate surroundings, thus either delaying or reducing any integration with their host culture? Or is it only a minor theme in the overall process of integration, perhaps slowing but not really altering the process of either assimilating into subcultures (such as Irish-American) or forging new sub-cultures? Will people juggle multiple national 'identities' of home and host society? Perhaps the realities of everyday life will take precedence over a 'long-distance nationalism' supported by technologically mediated communication. After all, however effectively ICTs can maintain contact with events, friends and relations, it is still impossible to replicate the many facets of being at home. One cannot replicate chance contacts with acquaintances and friends of friends. Happenstance and serendipitous events cannot be recreated through the intentional, narrowly defined, confines of computer-mediated communication. In addition, many face-to-face physical events are multi-modal – it is one thing to read a newspaper report, but equally important are the subsequent discussions with friends or co-workers, followed up by a television report. Since technological access to one's home culture (including friends and relations) is a series of isolated, rather multiple reinforcing, experiences, there may be significant restrictions on the long-term impact of such access. The short answer to this question is that no one knows; the data are, as of yet, insufficient to be certain, especially in the face of rapid technological change. However, with every year, electronic communication mimics face-to-face communication more effectively, thus continually changing the constraints on culture and community.

Emigrants may also create an identity that is based on the home culture, yet distinct from it. Expatriates can maintain contact with each other, and fashion an identity that is no longer fixed or constrained by geographical factors. The members of these groups are dispersed but have a defined national identity, and, as a group, they can engage in collective economic and political activity. It is well accepted that emigrant populations often act politically, on issues relevant to their home culture, and they use email, electronic discussion lists, and web pages to co-ordinate those actions. The most active electronic discussion lists in the world are lists whose members are expatriates of China, India and Pakistan, and expatriates from many other countries organise electronically to effect changes in their home countries (Anderson 1998). There is now scope for a national identity that is not limited by geographic boundaries. What will be the criteria for 'authentic' claims for citizenship rights? Is one a citizen of Ireland by being the offspring of an Irish citizen, even if born in a different culture? Is citizenship an entitlement only of those who participate in the Irish state (e.g. pay taxes), and is it a permanent entitlement or dependent on continued participation? The days of simple criteria for state membership or national identity are over, especially as different states use different criteria.

Of course, in addition to national, ethnic, and religious identities, occupational identities also constitute a strong claim on loyalty. Not only do policemen or firefighters feel strong loyalties to other policeman and firefighters in their county, but this loyalty can span countries and override national differences. This has often been illustrated in reactions to disasters. The reaction of firefighters throughout the world to the enormous loss of lives of New York City fire fighters after the 11 September tragedy is an example of such cross-national identity. Most individuals participate in a number of distinctly differing cultural or social groups or locations: the experiences of one's physical location; experiences of one's home culture; experiences of work-based groups. Individuals now choose which groups to participate in, do not have to make exclusive commitments to one group over another, and, inevitably, have less commitment to any single collectivity. Modern identity is multiple non-exclusive affiliations, rather than single exclusive affiliations, and affiliations determined by choice rather than birth. People often juggle multiple affiliations, many of which have little to do with nationalism or ethnicity.

10.3 Virtual communities?

New communications technologies enable individuals to work, shop, bank, and even socialise 'on-line', potentially resulting in less interaction with people and organisations in the locality (Nie 2001).[8] Does this undermine the fabric of local communities, as neighbourhoods become populated by strangers who no longer know one another and help each other? The perceived demise of community has been a concern for decades, long before the advent of the Internet and home banking (Wellman 1988), and the threatened demise of local communities may reflect industrial and post-industrial economic change, rather than the digital revolution (Putnam 2000).

A community is usually understood to be a group of people who share a common sense of 'belonging' (although belonging to what, and in what sense, is the complex issue). Individuals have multi-faceted interactions with others (Barnes 1969; Frankenberg 1969; Gluckman 1971), and their lives are embedded in a web of relations and commitments. Community is somehow an amalgam of the interactions, common experience and collective commitment among individuals who share long-term social relations. Often, these relations are involuntary as much as voluntary, forcing people to interact with people they do not necessarily like. Living in a community involves public interactions, often leading to unintentional and unexpected interactions with strangers and acquaintances.

These days, however, many proximate neighbours are no longer embedded in crosscutting networks of obligation and mutual assistance. For such people, there is no extended family in close proximity, and interactions with neighbours are fewer in number and more superficial in nature. This means that there are fewer people to provide reciprocity, assistance or reaffirmation. If you are a woman with a first time pregnancy, from whom do you get advice if there is a problem? There may be no neighbours whom one feels able to consult, and relations, from whom one would feel the right to claim assistance, may be geographically distant. In addition, people work in locations removed from where they live, and shopping, socialising and recreation may take place in different areas yet again. The advent of online access to services has accelerated this isolation from the locality, as people shop, bank, work and even view movies without having to interact with anyone else in their locality.

For some, the demise of local communities and the personal isolation and alienation linked with that demise have been solved by creating virtual communities. People use new technologies to find others with whom to share common experiences and concerns and, in so doing, create a complex web of interactions and experiences amongst themselves. These groups are composed of people who know each other, and help each other out, with reciprocal exchange. There is give and take or a barter system, with rules about how one behaves. The overriding notion is solidarity and that people put the interests of others and the interests of the group above self-interest, to create 'community', in this case, a 'virtual community'.

There are arguments which challenge the validity of the term 'virtual community'. Traditionally, communities implied geographical contiguity and face-to-face interactions, leading to overlapping, multiplex relations. People knew lots of different things about other people, rather than simply one facet of them. Participation in such communities was often involuntary: one lived in a locality and had little choice about who to interact with when out in public, nor could one be anonymously observing others without being seen. There were serendipitous meetings and unexpected events, there were public forums and private meetings. These characteristics contrast dramatically with electronic communities, where people only know one facet of others, participation is by choice, and people can sometimes 'lurk' without being observed. There is often no mechanism to verify that the person is actually who they claim to be and fake identities are easier to maintain in an electronic environment than a face-to-face environment. Critics argue that electronic communities lack the crucial elements of 'community' and are only 'ersatz' communities.

While virtual communities and proximate communities are different in many ways, this does not necessarily invalidate the existence of virtual communities. As with 'virtual reality', the descriptor 'virtual' in front of community does not mean that a virtual community must be a replica of a traditional community, but only that a virtual community encompasses the crucial elements of a traditional community. What are these elements? For some, face-to-face interactions over a long period of time are crucial, and these are, almost by definition, impossible in a virtual community. Despite improvements in technology, electronic communication does not mimic the rich, multi-modal communication of face-to-face communication. But is face-to-face communication a necessary or sufficient criterion of community? Everyday, face-to-face interactions are often superficial and narrow, and most people spend an increasing amount of time interacting with virtual strangers. Thirty-second interactions with the bus driver or the newspaper seller do not constitute interactions of social significance or meaning that create community.

Some of the other characteristics of traditional communities are long-term membership, multi-faceted interaction, mutual assistance, involuntary and unpredictable interactions, and a common set of practices or procedures that helps distinguish between members and non-members. Can any of these attributes be replicated in the electronic world, and, if so, would that replication constitute a community? In times past, the proximate community was a source of information and assistance. Such information and assistance can now be obtained via computer-mediated communication, and these interchanges can create a sense of common identity. If a child suffers from leukaemia and the parents do not know anyone in the area who is knowledgeable, or if they do not want to keep asking their local doctor (whose knowledge of the specialised subject may be superficial, in any event), they can search for information using the Internet. Often, the information may be available but difficult to evaluate or interpret. The most effective solution is to join a discussion list of other parents in a similar situation: they will have learned the spec ialised knowledge and will 'repackage' it in a way that the novice parent can understand. Out of this will often develop a community of people, who share knowledge and advice about this special topic. There is a support that comes from sharing experiences with others in the same situation, as well as the pragmatic benefit of sharing solutions to common problems with others in the same situation.

Discussion lists develop to share information on a variety of topics. These discussions groups are communities of interest. Members share a common interest, either in terms of leisure activities (such as football), work activities (such as a particular computer programming language), or personal circumstances (pregnancy, illness, handicap), and they use new technologies to share information and even organise collective activities. Are such groups communities? They may have long-term memberships, an internal social structure, and rules about how people communicate with each other. Most importantly, a sense of common membership and identity develops over time. People develop shared understandings through the course of interaction and communication that resembles a 'community of practice' (Wenger 1998): 'an aggregate of people who come together around mutual engagement in an endeavor [and] . . . practices emerge in the course of this mutual endeavor' (Eckert and McConnell-Ginet 1992: 89–99). Out of these practices comes a subjective experience of the boundaries between members' community and other communities and a sense of common identity. Some diagnostic characteristics of such 'communities' are:

- rapid flow of information and propagation of innovation
- absence of introductory preambles and very quick setup of a problem to be discussed
- substantial overlap in participants' descriptions of who belong and mutually defined identities
- specific tools, representations and other artefacts, shared stories and inside jokes
- jargon and shortcuts to communication
- a shared discourse that reflects a certain perspective on the world. (Wenger 1998: 125–6)

Communities of interest may have fluctuating memberships and fluid boundaries, but members share a sense of common membership and share a cognitive system focused on practices and rules. The members create the shared social system that is one possible criterion of 'community'.

It is sometimes argued that electronic communication is not 'real' communication, so 'real' communities cannot develop out of electronic communication. After all, electronic communication is a limited mode of communication, which excludes the communication channels (e.g. non-verbal) that are necessary for 'real' communication. Studies of computer-mediated communication from as early as 1984 (Kiesler, Siegel et al. 1984), and continuing since, often focus on the lack of cues and social context, and the diminished 'social presence' afforded by technologically mediated communication (for recent research, see Watt, Lea et al. 2002). The limits of computer-mediated communication are assumed to constrain electronic social relationships, and, since it is impossible for 'real' bonds of friendship and sharing to develop, it is also impossible for communities to develop.

Ethnographic evidence shows this is not correct. Members of electronic groups can develop strong emotional links with one another, and this commitment can include emotional support and mutual assistance, or may be defined in terms of common commitment to a collective ideal. Indeed, the bonds amongst members can become strong enough to mimic the moral obligations associated with family and kinship. Case studies, based on numerous ethnographic and biographical accounts, provide clear evidence of such collective commitment in a virtual environment (e.g. Rheingold 1994; Baym 1995), even amongst individuals who have no prior, or parallel, face-to-face communication. In the early days of many electronic networks (such as Usenet, Bitnet, World Wide Web), individuals co-operated voluntarily for the achievement of common goals. Individuals expended long hours for very little personal return, sharing a commitment to common goals and ideology (Hafner and Lyon 1996). The same can also be said of many community electronic networks and bulletin boards. Such groups are similar to voluntary groups that exist within industrial societies: sports clubs, religious associations, and neighbourhood assistance schemes (see Kollock 1999 for an example involving large-scale co-operation). There are also support groups for a variety of illnesses and disabilities, and individuals frequently report experiences of solidarity and mutual support with electronic groups (see Rheingold 1994; Rheingold 1993 for such accounts; as well as Nettleton, Pleace et al. 2002). There are also public interest groups whose members communicate electronically (such as environmental groups), and may share a strong collective loyalty to each other. All of these demonstrate a strong emotional commitment amongst people whose sole, or primary, mode of interaction is electronic.

Large-scale surveys of Internet use in the United States suggest that virtual communities are becoming an integral part of social life. A survey (PEW Internet and American Life Project 2001b) reported that 84 per cent of all Internet users had contacted an online group at one time or another, and 79 per cent of those identified a particular group with which they remained in contact. About one quarter of this group exchanged email with other members of the same 'community' several times a week, and half of them reported the main reason for contact was to 'create or maintain personal relationships with members'. Many of the reasons offered for participation in these communities are similar to the reasons that might be offered for participation in face-to-face groups: discussing issues with others and creating and maintaining personal relationships with other group members, discussing issues affecting the group, building relationships with others in the group. Some of these findings are open to question. For instance, it is not clear how strongly committed such people actually are to these virtual groups. One year earlier, the same research group reported that only five per cent of those with online access participated in a chat room or in an online discussion on a average day, although 28 per cent reported occasionally participating in a chat room or in an online discussion (PEW Internet and American Life Project 2000). Nonetheless, the trend, at least in the United States, is towards the emergence of collective, community-like, relations amongst people who communicate electronically.

This is not to say that people who communicate electronically always develop such strong loyalties, nor that communities of interest always develop into groups dominated by reciprocity and collective commitment, but it is clear that such loyalties are a possible outcome of such electronic communications. After all, groups whose members communicate by means of face-to-face communication do not necessarily develop a strong sense of mutual loyalty; traditional face-to-face communities are often dominated by division and conflict, held together only by the necessity of co-residence. Whether one looks at groups whose members communicate face-to-face, electronically, or both, one must not confuse content of communication with mode of communication. It is clear that a wide spectrum of social relations can develop amongst members of a group, regardless of the mode of communication. Individuals who reside in the same locality and communicate face-to-face may have little in common with other (other than common residence), or they may share ties of reciprocity, mutual assistance and loyalty. Equally, the same variation can exist amongst groups whose members primarily communicate electronically. Virtual communities may consist of people who interact to share information, or people who are devoted to some collective purpose, or people who provide mutual support and assistance. Such varieties of virtual community are becoming increasingly important as another means by which individuals create bonds with others.

10.4 Community networks

If industrial development over the past century has undermined the viability of local communities, the information revolution seems to have hastened the demise of such communities. Much of the social life of local communities depended on individuals leaving their homes to carry out such economic activities as shopping or going to the post office, The consequences of such activities were unintentional (sometimes) and often unavoidable interactions with friends and strangers. Going to the shop, one saw a neighbour and exchanged news. After going to the shop, one might have arranged to meet a friend for tea. Such activities helped create the sense of common knowledge, experience and commitment that sustained a sense of community. If economic – and even social – activities can be carried out from the home, then where does the basis for collective experience and identification come from? Will the interactions and shared experiences that we think of as 'community' shrink to the point where local communities cease to be anything more than physical aggregations of isolated individuals? While new technologies seem to undermine local communities at the expense of virtual communities, can they also enhance local communities? Does community life necessarily atrophy when people use new communication technologies and, if this is an undesirable thing, can the process be either arrested or altered?

It is certainly possible to use new technologies to supplement, complement, and, if necessary, replace traditional relationships and communications patterns in localities. If the information functions previously carried out via town meetings, bulletin boards in shops, church attendance, or other traditional patterns of information dissemination are no longer effective, new technologies can provide the same benefits in the local community. This can be done by providing an infrastructural resource for the provision of information. Thus, if the disappearance of the local shop also meant the disappearance of the local noticeboard which had helped to organise babysitting or the selling of second-hand furniture, then an electronic noticeboard can act as a replacement. If the local shop has not closed, but people do not have time to visit it, then an electronic noticeboard can supplement the physical noticeboard. If discussion of local issues has decreased because people do not have time to meet together, then an electronic discussion list, which people can access and contribute to when it suits them, can complement face-to-face discussions.

There are some intriguing examples of networks of neighbourly trust, and evidence of residents' desire to recreate such networks. For instance, a woman named Angie Hicks set up Angie's list in 1995 in Columbus Ohio. She gathered together a group of friends and family and started a list of good and bad service companies. It soon spread to 13 different markets, including areas in Ohio, Minnesota, Florida, Wisconsin, North Carolina, Massachusetts and Illinois. It attracts homeowners who can rate companies in 250 categories. For a $35 annual subscription fee, they can sign up to the list, and either provide ratings for services or check ratings that others have provided. The list has ratings on more than 10,000 service companies, and has 50,000 members. Members can view companies online or they can call 'neighbourhood specialists' who work in call centres (Power 2001). A similar, but perhaps less 'trustworthy', service is offered by epinions.com, which allows consumers to recommend products, and also offers them the chance to shop online. It has about one million reviews and comments. These are both examples of new technologies solving a problem that has always existed in communities (e.g. where can you find a good plumber or electrician?), which cannot now be solved by people who are isolated from their neighbours.

The effectiveness of new technologies in sustaining local communities is difficult to evaluate because relatively few localities have sufficient density of computer users and information providers. In so far as such community networks have developed, they have often been introduced to increase political participation and activism (see chapter 7), with an assumption that strong civic participation will also strengthen internal community structures. The Freenet movement in the 1980s was the highest profile of these experiments, first starting in Cleveland, Ohio in the United States and spreading throughout the United States and Canada, with some similar projects in Europe (Graham and Marvin 1996). Other experiments have had a commercial focus, as organisations tried to see if the public would be willing to use, and perhaps even pay for, various information or entertainment services (Dutton, Blumler et al. 1987). In the last ten years, there has been a substantial growth in the number of towns and cities providing information about services and local activities, often linked with government support for such programmes. Governments often support community networks in order to encourage community development, sometimes as part of a wider programme for rural development or social inclusion (Haase and Pratschke 2003; McCaffrey, C. 2003; O'Donnell, McQuillan et al. 2003; Loader 1998).

While the FreeNet project receives little government support in the United States, in other countries governments have provided more direct support. In Canada, there has been a 'Connecting Communities' to provide community access (http://www.connect.gc.ca/, Birdsall 2000), while New Zealand announced a similarly named programme in 2002 (http://www.dol.govt.nz/cegccstrategy.asp, but see also Crump and McIlroy 2003). Similar projects have been undertaken in the United Kingdom as well (see, for example, Communities Online, http://www.communities. org.uk/). Some examples include Grimethorpe Electronic Village Hall (www. barnsley.org.uk) to help unemployed miners after the closure of the local colliery in 1995 as well as the Warwickshire Rural Enterprise Network (www. nrec.org.uk/wren) to help deal with rural isolation. Similar projects have taken place in urban areas as well, such as the Asian Community Centre, Chorlton Workshop and Woman's Electronic Village Hall, all in Manchester (Agar, Green et al. 2002; Carter 1997).

The evidence regarding the impact of technology on local community life is ambiguous, with some evidence suggesting that new technology supplements local communities and face-to-face social life while other evidence suggests that it undermines both (see Kavanaugh and Patterson 2002 for a review of issues; as well as Haythornthwaite and Wellman 2001; PEW Internet and American Life Project 2001b). Not only do different studies show different results, but the same data have also been open to conflicting interpretations (Nie 2001). The Pew Internet and American Life Project (2002), while positive about the use of the Internet for 'virtual communities', is less positive about its benefit for local communities. The greatest use of the

Internet, in terms of local communities, is local information: four out of ten Internet users report that they 'often' or 'sometimes' go online to look for information about local stores or merchants and about one out of three Internet users look for news about their local community or information about community events. These are community 'replacement' benefits from Internet use, but the report finds little evidence for greater participation in local policy issues.

Local community networks require local information providers as well as consumers. Significant motivation from a large number of participants is needed to provide relevant information, as well as a strong local interest in accessing such information. For the moment, most community networks are single sites, on which various local services are flagged and local events announced; these sites depend on specialist workers or volunteers. In some cases, service providers (plumbers, sports clubs) may also have their own sites, linked to the main site, but the majority of residents are not involved. This differs from the communication pattern of face-to-face local communities, and so the development of such local community networks has been slow. It remains unclear, if barriers were absent and density of usage was sufficient, whether the social, economic and political transformations discussed would actually take place. That is, if people were able to choose an electronic, wired life, would they? If they did choose such a life, how would that alter their everyday face-to-face life? Would one develop, to the cost of the other, would the two worlds exist side by side, or would the two worlds intermingle?

One detailed study of the impact of technology on community social life has been carried out in a Toronto suburb where, from 1997 to 1999, residents in a newly built estate of about 120 homes were provided with high-speed Internet access, and the relevant computer technology with which to access information and entertainment and to communicate with others. This was done by a consortium of companies who wished to examine commercial potentials of such 'wired communities' (Hampton and Wellman 2000). About one third of the residents either did not, or could not, participate in these high-speed trials, so it was possible to compare 'wired' residents with unwired residents. Based on surveys and examinations of data traffic, wired residents 'recognise almost three times as many neighbours, talk with nearly twice as many, and have been invited, and have invited, one and half times as many neighbours into their homes in comparison with their non-wired counterparts' (p. 205).

Similar results were also found in a study of the Blacksburg Electronic Village in Blacksburg, Virginia in the United States (Kavanaugh and Patterson 2002). The study found evidence of community involvement linked with new technology, but was unable to determine whether it was people already involved in the community who were now were using the new technology or whether the new technology fostered more community involvement.

The Blacksburg project was heavily supported by a local university, and the impact of that university may have distorted the relationship between technology and community development. However, evidence suggests that if community involvement is already strong, new technologies can maintain that involvement.

A notable example of community-based Internet and computer use has been in Ireland. In 1996, Telecom Éireann (the Irish telephone company) held a competition for an 'Information Age Town'. The company would 'wire' a community, providing inexpensive computers and telecommunications access, communications infrastructure, as well as training for both private individuals and commercial organisations. Towns with a population of between 5,000 and 30,000 inhabitants were invited to submit development proposals, detailing how they would use the £15 m. investment for the benefit of the town. From the company's perspective, the investment was intended to answer a number of questions:

1. What happens when every home has a telephone – not just an ordinary telephone, but one with sophisticated voice-mail, caller-line identification and other advanced services?
2. What happens when every business, large and small, has access to an ISDN connection and high-speed access to the Internet?
3. What happens when every student in the education system, from the age of five, has regular, intensive access to a computer with learning, knowledge-gathering and communications tools?
4. What happens when public services – from libraries to healthcare – are fully equipped to exploit the potential of the Information Age?
5. What happens when the majority of households have a personal computer linked to the Internet? (McQuillan 2000)

The motivation was to develop a test bed for the provision of various electronic commerce services, as well as other telecommunications intensive services, to see what the commercial future of a telecommunications provider might look like.[9] Unlike the Blacksburg project, there was no direct involvement by a university, and technology intervention from external agencies provided only training and support in the initial phase.

The winner of the competition was Ennis, a town in the West of Ireland with a population of about 16,000. The investment involved the provision of subsidised computers for home use for those who qualified. Subsidised PCs were offered to 5,600 residents. Residents were required to contribute about 15 per cent towards the multimedia computer plus software package, and a free PC familiarisation plus one year's free Internet connection was offered. Eighty per cent of the residents qualified for the subsidy, and 85 per cent of those who qualified, then took advantage of this offer. As a result, in the first survey (McQuillan 2000), the results showed that Ennis had the highest household rate of computer ownership in Europe:

- 83 per cent of households own Internet enabled, multimedia computers
- 91 per cent of these households activated an Internet account
- 45 per cent of Ennis households have at least one active Internet user

By the second survey, in June 2001, levels of PC and Internet access were more than twice the national Irish average (household penetration of PC's of about 80 per cent, whereas the national average was 32 per cent). Interestingly, of those Ennis residents who had a computer, 76 per cent had an Internet connection; while 63 per cent of those with a computer nationally had an Internet connection, The major difference between Ennis and the rest of Ireland was the rate of computer ownership (although the proportion of computer owners with Internet access was similar), so the first lesson from Ennis would seem to be that if provided with cheap computers and training, a significant number of people will use them to access the Internet.

As with studies from the United States, the largest volume of Internet usage has been to maintain contact with friends and relatives, with 78 per cent of Internet users maintaining contacts, on an average of eight times per week. Email did not replace other modes of communication; while email was used to maintain social contacts by most Internet users in Ennis, the Internet accounted for only 17 per cent of actual contacts in this area. The balance of social contacts was by such traditional means as phone, fax or letter. This may suggest the emergence of a wired community, but not everyone uses the wires. Amongst houses with computers, almost all used their PCs, and accessed the Internet, at least occasionally, but only one in four used the PC/Internet on a daily basis. In the first survey in 2000, over 55 per cent of households had no Internet user in the house, despite inexpensive access to training, technology and telecommunications. A substantial number of people did not find the Internet useful in their daily lives, at least in the initial phase of the project. One year later, while the reported daily usage was about one in four, over one-third of individuals used the Internet once a month or less.

The impact on local organisations of wiring this community seems to have been mixed. There are some examples of community organisations using new technologies to at least maintain, and perhaps improve, local participation. However, on the Ennis website, while there are 33 organisations, only some of them host their own websites with information on current activities. The evidence is still slight, but it appears that, as with the Blacksburg project, new technologies enable existing community activities in Ennis to flourish but it is not clear that new community activities follow from the introduction of new technologies. More evidence from long-term and in-depth community studies throughout the world is essential, but early data suggest that new technologies facilitate those who are committed to maintaining community involvement. However, there seems little evidence that the introduction of such technologies, of themselves, ensure the continuation of community life in rural or urban areas. It appears that a revival of community life and increased 'social capital' depend on more than an injection of information and communications technologies (Putnam 2000; see also Wellman, Quan-Haase et al. 2003; Prell 2003; Gurstein 2003).

One of the themes that often underpins discussions of community networks is that if people are not using community networks, it must just be a matter of demonstrating these benefits to people or reducing the barriers and adoption will follow. Yet one of the findings from the Ennis project was that even when there were no barriers to use (the computers were available cheaply and easily, training was free, and telecommunications charges were subsidised), a sizeable percentage of people still have no interest in using the Internet to access local information or to 'surf the web'. This conclusion is not unique to the Irish experience, as it has been demonstrated by an increasing number of other studies (Wyatt, Thomas et al. 2002; Ward 2000; Lenhart, Horrigan et al. 2003). A significant number of people have no desire to access the Internet or participate in community networking. This is not because of fear, ignorance, high costs, or economic disadvantage; they simply do not see it as relevant.

However, 'opting out' in this way may not be viable in the long run. As technology becomes more intrusive and pervasive, people may have no choice but to adopt the technology. For instance, when the bank reduces staff costs by closing the local branch office and national governments do the same for the local post office, access to cash may become possible only via a automated teller machine and banking transactions may require Internet access. In which case, people would have little choice but to use banks cards and the Internet. It is an increasingly common complaint that people are being forced to use technology because they can no longer avoid it. This is not inevitable. In the case of the local bank branch and post office, governments could subsidise a service which might be uneconomic but which citizens preferred. Should individuals have to adopt technologies that primarily suit the needs of governments or corporations, or should the technologies be designed also to provide perceived benefits to individuals? There is scope for policy intervention by governments on such issues, and such interventions may be the prime determinant of the future of community networks.

This mix between individual choice and collective policy in the future of community networks is symptomatic of all the changes discussed in this chapter. The impact of the information revolution on culture, identity, and community will be a combination of individual choices, collective policies and economic forces. This will be a future that varies from one community to the next and from one society to the next, remaining mutable and pliable.

Chapter 11

Reprise

Whether twenty-first century societies are described as information societies, post-industrial societies or advanced industrial societies, there is little disagreement that economic, social, political, and cultural changes are taking place. The starting point for such change must be the computer revolution and digitalised information. Developments in computer and telecommunications technologies have led to the proliferation of cheaper, faster, and more powerful computers that are available to an increased number of people and used in increasingly diverse ways. More information, and more types of information, can now be encoded in digital, rather an analogue, form. The inexpensive yet accurate reproduction of information makes information accessible to all. With the emergence of a digital communication infrastructure, this information can also be inexpensively and quickly distributed, to single or multiple individuals, potentially creating wide audiences for individual information publishers or producers. This digital revolution has meant an increase in the amount of information and an increase in its range of circulation. People have access to information that previously would have been restricted to a relatively small audience. There has been, quite literally, an information explosion.

Will everyone participate in this revolution? In the early days of the Internet, users tended to be young, male, often working in technology-based organisations, who lived in the United States. Such users are clearly atypical, and it was difficult to know whether the technology would become widespread or whether the use of the technology by such an unrepresentative group could be the basis for extrapolating to the wider population. Recent research indicates that many of the disparities that characterised the beginning of the online revolution are disappearing. The gender gap has disappeared (women as likely as men to use the Internet),[1] the urban/rural divide is disappearing, and the age divide is evening out. The young, especially the 15–20 year old age group, are still the highest Internet users, but use is increasing for users up to and including the 50-year range. Individuals living in low-income households or having little education still trail the national average in the US, but growth continues to be most rapid amongst those who use Internet least, providing some hope that gaps will narrow over time (National Telecommunications and Information Administration 2002). In September 2001, 54 per cent of individuals over the age of three years had used the Internet (National

Telecommunications and Information Administration 2002), and survey evidence suggests that in 2002, the number had risen to 60 per cent (http://www.nua.com/surveys/). Future trends would suggest increases in all of these areas; the use of high-speed, broadband Internet access is increasing; in the year between 2000 and 2001, high-speed access doubled from ten per cent of Internet users to twenty per cent, with every indication of greater increases in the future. This will encourage even more – and more varied – Internet usage. It seems that the digital revolution will eventually be part of everyone's life in the United States, regardless of region, class, ethnicity, gender or education, just as the telephone or television.

While computers and the Internet in the early days of this revolution were the preserve of the United States, any perception that the United States is the still the most technologically advanced society is exaggerated. In an International Telecommunication Union report which ranked countries according to digital access, Sweden, Denmark, Iceland, Republic of Korea, Norway, The Netherlands, Hong Kong, China, Finland and Canada were all ranked more highly than the United States (International Telecommunication Union and Minges 2003). Other surveys, using different criteria, might arrive at different rankings, but it is clear that the United States is only one among a number of societies which have been permeated by new technologies. Even though many societies would still rank low on any index of information and communication technologies, the trend in all societies is for a greater and more even distribution of technologies.

New technologies have enabled co-ordination of activities and exchange of data, so that companies and corporations now exist in a global economic and technological environment, with constituent parts of corporations distributed throughout the world. The introduction of new technology has led to more rapid manufacturing of products for niche markets, assisting a consumer revolution. New technology has been responsible for the emergence of a new information industry as well as the transformation of traditional economic sectors. Information provides 'value-added' content for all activities ranging from farming to industrial production to service industries, often determining the profit and efficiency of any of these activities. Growth in the technological infrastructure has led to a global and distributed economy, with economic interdependence amongst countries throughout the world.

These economic changes are so far reaching that a return to an industrial economy or the kind of society linked to an industrial economy is no more possible than a return to agricultural production or an agrarian society was possible after the industrial revolution. The clock cannot be turned back; we now live in an information dependent society that is a complex of interdependent parts and is quite fragile. If the information is wrong or if the information technology breaks down, chaos results (e.g. Peltu, MacKenzie et al. 1996). Information, as a whole, is a significant aspect of our lives, even though it is becoming more difficult to trust or verify individual 'bits' of information. Individuals, and society as a whole, are dependent on information and the co-ordination that new communications technologies enable. The diffusion of information technology throughout society may be slow and uneven but it is pervasive. Organisations are incorporating new technologies into their activities to improve productivity. Governments are incorporating new technologies to permit more efficient and effective delivery of services. Individuals are incorporating technology into their everyday lives to make their lives easier and to adapt to a changing society and economy.

This seems to paint a picture of inevitable technological change, but such a picture is dangerous because it encourages a belief that economic and technological changes are external to individuals and to society.[2] It would be as though we were all bystanders, watching a drama unfold that followed an inevitable script, while we tried to guess the next piece of dialogue. The fallacy that individuals have no choice in the process taking place is a dangerous aspect of this technological revolution. It is as though technical and economic change, like the *Titanic*, is unstoppable and we can only look out to see if there are icebergs ahead. But, in fact, if there are icebergs, the ship can be turned. Technological and economic changes are driven and directed by broader social, political and cultural pressures, but this is often not apparent. Individuals and groups can, if they wish, exercise choice in the process of technological and economic change through individual choices and government policy decisions.

A global information economy is emerging, but it is a differentiated economy with much room for local variation. Different locations and societies will participate in this differentiated global system in different ways. Will Ireland provide relatively cheap labour for a world economy, or will it be able to provide high paid employment based on education and skills? If the policy is to be the latter, will governments provide the necessary funding for education and research for such skilled labour, and will the technical and legal infrastructure be provided to support such activities? On what basis will governments decide on centralised versus decentralised economic growth (Government of Ireland 2002b)? Will government policy encourage employment in clustered regional areas by evaluating which economic activities are suitable for decentralisation and providing appropriate support for such activities? Global economic change seems inevitable, but different outcomes are possible for different segments of this global information economy. There is scope for policy interventions to influence the place of particular societies in this global future.

There are similar economic choices to be made regarding individuals. New technologies have reduced the number of personnel necessary for production in agriculture, manufacturing and service industries, and made the labour that is still needed both cheap and easily replaceable. What will be the future for such people? Will there be people providing socially beneficial, even if not economically necessary, services? If so, who will pay them? Alternatively, if there is no work, will a permanent underclass emerge, or will people be provided with enough resources to lead satisfying lives? For those that remain in work, it is clear that work conditions will be altered by new technologies. Will work be subject to surveillance and minimal regulation, or will work conditions be protected by the state? Will work become part-time, temporary and poorly paid for most, while remaining full-time and well paid for a few? Will individuals find themselves having to use new technologies in their daily lives, regardless of their own preferences? There are policy decisions that can be made by citizens and governments that will influence such alternative outcomes.

The emergence of digital information has also transformed the social impact of information. The result of the digital information revolution has been an information explosion, but the cost of this information explosion has been a decrease in the relevance and accuracy of information. The cost of information production, distribution and consumption formerly restricted the amount of information that could be published. As a consequence, editors and publishers filtered the information that was available to individuals. These mediators found information, evaluated its accuracy and determined its relevance for an audience, excluding information that was inaccurate or irrelevant. Such limitations no longer restrict information production; individuals must now find and judge the relevance and accuracy of information themselves, in the context of exponential increases in the amount of information available and decreases in the authenticity of information.

Individuals are now more dependent than ever before on information, and we intuitively recognise this information dependence by our concern with information overload. There is too much information, but we do not dare ignore the phone calls, electronic mail messages, postal deliveries, and voice mail messages because there is the one piece of information in a hundred or thousand that might be accurate, relevant and crucial. Since we have no way of filtering accurate from inaccurate, relevant from irrelevant, or crucial from unimportant, we have no choice but to look at each piece of information in turn or become dependent on filtering programs devised by others. There also remain those whose lives are increasingly controlled and constrained by information, but have no access to information themselves. Will there be policy initiatives to provide access to digital information, as there have been policies to provide access to print information?

Technology has changed communication patterns as well. Electronic mail, mobile phones, text messaging, answering machines and voicemail, instant messaging and other 'real-time' communications programmes are all part of many people's everyday lives. These are technological solutions to the problems of everyday life; daily life is more complex and unpredictable, and we now take for granted the micro-management of daily life, as new technologies allow us to co-ordinate activities and change schedules. Social lives are fluid, and rearranged on the basis of contacts with multiple people, using multiple technologies, in short spaces of time. Those who do not have the time to shop in person can shop online. Those whose job takes them away from friends and relations can use electronic mail and mobile phones to keep in touch. The list goes on. However, many of these technologies are solving problems that have been created in the first place by technologically driven economic change. Electronic mail may help maintain contact with friends and family scattered throughout the world, but they are scattered because of the global economy that can only exist because of those new technologies in the first place. Mobile phones help co-ordinate our complex daily lives, but these lives are complex because of the demands of a modern economy created by new technologies. These are all user driven adaptations, as we embrace technology to solve problems that have resulted from technology.

The digital revolution has meant a loss of privacy, sometimes without individuals being aware of it. We now live in a more public world, where more aspects of ourselves are available for inspection by others whether we like it or not. It is now a simple technical process to capture increasing amounts of information about individuals' lives, and privacy is possible only if appropriate policies are in place. It is important to control the amount of private information collected, but there are many aspects of a person's life that cannot be private if society is to function effectively. For instance, financial information has to be available if credit card purchases are going to be verified; the only alternative is not to use credit cards. Because information collection is inevitable, there must be regulations about who has access to information, how long the information is kept, limits about what that information can be used for, and other measures to protect privacy.

To some extent, these communication technologies are recreating, for a geographically dispersed population, the level of social interaction that would have previously existed in rural, and even urban, communities. Instead of contacting a relation or friend in the normal course of life in a small community, new technologies enable us to stay in contact and participate in the lives of people in distant locations. The personal consequences of these technologies have sometimes been subtle, but are often far reaching. Communication is now person based, rather than location based, and people are in contact with each other regardless of where they are or how often they have moved. Frequently, the level and amount (though not necessarily the significance) of the communication have increased dramatically. However, technologically mediated communication differs from face-to-face communication, and a social relationship based on electronic mail and telephone conversations will be a different relationship from that based on face-to-face relations. This is not only because of differences in information content, but also because of differences in the context of interaction. Electronic communication is voluntary and intentional, as well as limited in content, whereas face-to-face communication is both voluntary and involuntary, intentional and unintentional; these differences alter the nature of social interaction. This parallels the difference between traditional face-to-face communities and electronic communities. Electronic communities are voluntary communities whereas in face-to-face traditional communities, inhabitants had no choice but to interact with each other. In electronic communities, people often simply stop participating (effectively, 'leaving' the community) when confronted with difference or conflict (Komito 1998a; 2001). When technologically mediated communication makes it easy to avoid conflict, will people become less able to resolve conflict?

Will individuals also become more isolated from their local neighbours? New technologies have enabled individuals to carry out economic tasks while diminishing the need for social contact. As economic life is disembedded from the social context of exchange, this further accelerates a process that began with the industrial revolution. This potential demise of local communities as people cease their participation has inevitable implications for individual isolation and alienation, and requires appropriate social policies to address such tendencies. On the other hand, new technologies also offer the possibility of encountering and communicating with new people. Surveys of Internet usage make it clear that people are 'meeting' new people and making friendships. These people share interests or concerns that are not shared by face-to-face neighbours, and new technologies enable such people to overcome both isolation and alienation.

In the new economy, cultural commodities, such as food, clothing and movies circulate globally and are available for local consumption. Will the unique experiences and social relations of locality diminish, to be replaced by a uniform global culture? Will the same architectural styles, eating habits, and clothing fads exist everywhere, just as every international hotel now offers the same standard 'international breakfast' menu[3] and Christmas celebrations look the same anywhere in the world where Christmas is celebrated?[4] If people have fewer shared experiences with their neighbours, will participation and identification with local groups, whether sporting clubs or neighbourhood help, also become a thing of the past? Experience and identity are now an issue, as we juggle local and the global, as well as face-to-face and electronic communication.

New communications technologies may seem to encourage global uniformity at the expense of local identity, but there are also countervailing pressures towards diversity, especially 'cyber-diversity'. There used to be a relatively uniform or homogeneous information universe: everyone received the same information, which was collectively accepted as relevant and accurate. Increased amounts of information mean increased information diversity, and there is now conflicting and contradictory information available from diverse sources. Friends, neighbours, kin and even family watch different television programmes, read different books, listen to different music, follow different clothing trends, prefer different foods. People in the same house, residing in the same street, or living in the same community, can live distinct and separate lives. This encourages segmented and compartmentalised sets of cultures, as individuals can be members of distinct groups, whose beliefs and values are reinforced by receiving only that information which is relevant to that distinct world. Will this lead to a society composed of diverse and mutual opposing groups, each with decreasing empathy and understanding of others' views? With the diversity of local experience, common global experience, non-territorial collective experience, and individual experience, what kind of societies will emerge, as people juggle face-to-face life, global commodities and electronic diversity?

These are all examples of policy decisions to be made about alternative futures, but there are no inevitable outcomes regarding any of these options. Policy interventions, guided by public deliberation, can impose a structure on the future. The future of any society can be partly determined by the policy decisions of governments, as well as the unintentional, but socially conditioned, choices made by individuals. It is not acceptable for governments to decide that new technologies will benefit citizens, and then force citizens to accept both the benefits and costs of those technologies. Nor is it necessary. People need not change in order to adapt to new technologies; technologies can be created or altered that will be appropriate to the needs and desires of citizens. This debate cannot be postponed. The alternative is to be a passive audience, watching a play that is written and performed by others, as societies integrate into the global information economy. The future is unlikely to be recognisable as a continuation of the past – it is not 'business as usual despite alternations', it is 'new premises under construction'. Furthermore, each society (or 'premises', to continue the analogy) will be slightly different. The challenge is to understand the changes taking place and decide what the future structure of society should be like. With that knowledge, it is possible to intervene and so exercise control over technology, rather than walking backwards into the future.

Notes

Preface

1 See Webster (2002) for discussion of these opposing views.
2 There are many such texts, some of them comprehensive and articulate (for instance, Castells 1998; 1997; 1996; Robins and Webster 1999; Webster 2002; May 2002; Stehr 1994; Beniger 1986; Dutton 1996; 1999; Roszak 1994 are all interesting and sometimes conflicting expositions). For a collection of central readings, see Webster, Blom et al. (2003).

Chapter 1 Introduction

1 This view was first popularised by Daniel Bell (1973).
2 Quoted in Navasky (1996: 216).
3 Debate about VHS vesus BetaMax has been going on for years and it is now one of many 'urban legends' (see http://www.urbanlegends.com/products/beta_vs_vhs.html). Liebowitz (2002) was the first academic to argue that VHS was not worse technically than BetaMax; the victory of VHS was largely due to marketing and user consumption issues.
4 See Corcoran (1993) for a study of Irish illegals in the United States.

Chapter 2 History of information

1 This is clear from any texts on human evolution, see Young (1971), or any standard texts on physical anthropology and human evolution.
2 Written text is also interpreted by reader, and the last decade has seen broad acceptance that written texts are also social constructions. Despite this, written text is still less subject to alteration than oral speech.
3 See Bernstein (1964) on restricted versus elaborated codes, as well as Douglas (1973) on cultural aspects of this distinction.
4 Not all writing suffers from such constraints; for instance, letters between friends are written in an informal style. The reader is already known to the writer and the two share a context by which ambiguous phrases can be deciphered. Equally, it is possible to write texts using a specialised vocabulary so that only other members of the same group can understand the written text. Such exclusionary practices are a way of recreating, in written form, the special relationship between speaker and audience in oral speech. Most written text, however, is intended for a wider, public audience.
5 The relation between science, analysis, and written text is a complex one. Having stated that science is only possible once facts are extracted from their social context, social studies of science and technology have, over the past few decades, demonstrated that such a detachment is more of an ideal than a reality. The claim that scientific 'facts' exist outside society is just that – a claim. 'Facts' exist

within a social context and the process of detaching 'fact' from 'person' only disguises this dependence on context.

6 Oral language can also be used in this way, in which restricted knowledge or vocabulary is used to create secret societies or clubs that exclude others from powerful knowledge. However, with written language and differential levels of literacy, such distinctions become much easier to institutionalise, so that a larger social group is included.

7 It also helped painting by making the mass reproduction of original artworks possible, again see Benjamin (1973).

8 This applies, of course, only to industrial societies where a telephone infrastructure is in place, as opposed to industrialising societies in which access to a telephone remains a preserve of the elite.

9 The only partial exceptions are the pirate or illegal radio stations that can sometimes flourish if there is sufficient demand, but these depend on low power transmission to a restricted audience. In such cases, it was affordable to simply set up new low power broadcast equipment, if equipment was confiscated or, alternatively, set up high power transmitters outside the jurisdiction. In both cases, it was virtually impossible to regulate content.

10 The only exception to this was the telegraph, which was rapid but also recorded. However, the amount of information was severely restricted, and the storage of information was temporary.

11 Although it is only a matter of time before both the production and consumption of visual and auditory signals for television and radio are also digital. Digital televisions are already becoming more common.

12 For further readings on the history of information processing technologies, see Winston (1998), Marvin (1988), Beniger (1986).

Chapter 3 Information technology and the digital revolution

1 This is a transformation from base ten notation, in which each column increases by ten (one, ten, 100, 1000), to base two notation, in which each column increases by two (one, two, four, eight, 16, and so on). The actual amount or number is the same, although the representation looks different. Thus 21 is represented, in base ten notation, by two in the 'ten' column which is added to a one in the 'one' column ('21'). The same amount is presented in base two by one in the '16' column, added to one in the 'four' column, and one in the 'one' column ('10101'). For more on this encoding issue, see Lynch (1974).

2 As late as the early 1980s, some universities in the United States refused to install word processing programmes on mainframe computers; word processing was seen as an expensive luxury for an expensive computer.

3 This is precisely the solution in some countries, where wireless computing enables countries to avoid the cost of installing telephone lines. This has enabled a mobile phone market to develop in Pakistan (Malik 2003), as well as Internet access in Afghanistan (Hammersley 2003). Users still have to be able to afford mobile phones or computers, as well as the cost of accessing the network, but at least the network costs are minimised by use of the new technology.

4 It is debatable whether the microcomputer revolution actually achieved this. Individuals now have access to significant computing power, and have access to information via the Internet, but organisations can still afford larger and faster computers, better software, and so on. The free information that is available to anyone tends to be out of date and unreliable, as compared with the information that organisations can afford to pay for. Further, information does not often necessarily convey power, sometimes it only reveals how powerless a person might be.

Chapter 4 An information economy

1 There has been considerable research showing that the economic consequences of new information and communication technologies are far more complex than can be elaborated here (see, for example, Stehr 1994; Dutton 1996; Castells 1996; Preston 2001).
2 For a discussion of changing wage levels in the United States, especially since the introduction of new technologies, see Moseley (1999).
3 See chapter 7 in Preston (2001) for a discussion of the impact of economic transformation on individual workers.
4 New technology can also be used to increase the amount of profit that such workers make, by enabling them to bypass middlemen and sell directly to consumers in more developed countries (see United Nations Conference on Trade and Development 2002).
5 In this example, calculating 'similar value' must take into a number of factors. These factors include whether the seller provides continual support, whether it requires new in-house expertise to use, whether it can be easily integrated with previously purchased software products, whether it is compatible with the products that other organisations use, and so on.

Chapter 5 An information society

1 http://www.sims.berkeley.edu/research/projects/how-much-info-2003/
2 This, needless to say, applies only to societies where the content of mass media is not subject to state control. In countries where the state determines information content, consumers learn very quickly to be dubious about the authenticity and accuracy of public information. Even the significance of the information can be in doubt, since there is likely to be a disparity between what the state and the public considers to be significant. There may even develop a 'black market' in alternative information.
3 There is a long established literature looking at the sociology of mass media and news reporting (e.g., Curran, Gurevitch et al. 1977; Pool 1983; Curran and Seaton 1991).
4 For further discussion of social studies of science and the sociology of information, see Merton (1973), Latour (1987), Yearley (1988), Webster (1991), and Gibbons, Limoges et al. (1994).
5 For discussion on the link between technology and the growth of the state, see Ellul (1964), Beniger (1986), Mumford (1934; 1962), Dandeker (1990), and Giddens (1985) to name a few.
6 With the exception of sacred knowledge which was restricted.

Chapter 6 Regulation of information

1 See Pye (1992) regarding the history of technology investment in Irish government departments.
2 In practice, since deciphering messages using public key encryption is computer-intensive and time-consuming, this system is usually used only to create and share a symmetric key which will be valid just for that session. Since the symmetric key is shared using a secure system and will only be used for a limited period of time, it is effectively private. In the case of World Wide Web browsers, for instance, the software uses public key encryption to create a symmetric key that is valid only for the duration of the secure web connection.
3 In fact, while there is evidence of the use of public and anonymous email services for co-ordinating illegal activities, the content of these email communications is not encrypted (Campbell

2001). Instead, the emails depend on messages that would be ambiguous to anyone that did not have a shared knowledge of the backgrounds of the individuals – a time honoured and effective way of talking in 'public' so that only one's own friends understand the vague references and descriptions. In any event, since the use of encrypted communication is so rare, use of this kind is more likely to reveal identities of such individuals, since encrypted communications are more easily identified through electronic surveillance than unencrypted but unclear communications.

4 Current laws and proposals in various jurisdictions are tracked by the Electronic Privacy Information Centre (http://www.epic.org).

5 This issue arose in the earlier discussion of court cases involving Microsoft and other software companies.

6 This is a constantly changing area, and the best source for current debates is the Electronic Frontier Foundation (http://www.eff.org) and Electronic Privacy Information Center (http://www.epic.org).

7 There were, of course, exceptions. Many consumers in Ireland could receive BBC and ITV without need for relay stations, if they lived close enough to line of sight broadcasters and had directional aerials.

8 For a general discussion of these issues, see European Commission Legal Advisory Board (1996) as well as the EU Green Paper on *The Protection of Minors and Human Dignity in Audiovisual and Information Services.* (http://europa.eu.int/ISPO/infosoc/legreg/docs/protect.html).

9 As reported by BBC, http://news.bbc.co.uk/hi/english/world/ europe/newsid_524000/524951.stm

10 Such countries are less likely to have high-speed international connections, which reduces access by the global public, and they may also have poor quality software and hardware support services.

11 For instance, the user can jump to a site outside France and then access the site from that remote location; the user can phone an ISP outside France, find a mirror of the site outside the United States, or make use of a host of other strategies to short circuit such a ban.

12 The Republic of China, for instance (as of 2001), controls ISPs and this enables the government to block access to sites if they so desire. Sites blocked often include the BBC site, perhaps because there are links from that site to both text and audio versions of its Chinese-language service (Gittings 2001).

13 In the wake of 11 September, new laws in the United States require librarians to keep track of which books are borrowed, using the same logic. Civil liberties groups, as well as librarians, are opposed to what is perceived to be an infringement of personal freedom (see http://www.ala.org for the American Library Association's response to the Patriot Act). The same monitoring is proposed for access to electronic information.

Chapter 7 Political participation

1 Although the Group also warned that 'the increase in the flow of information does not necessarily engender an amelioration of the democratic system. It could just as easily lead to a distancing of citizens with regard to real democratic stakes'.

2 Everyone, that is, that had a political voice. In many non-state societies, this excluded women and children and might exclude individuals without kinship links.

3 This process may also involve consulting ethnic or religious groups as well.

4 Howard Dean raised enough money to enable him to withdraw from the public campaign finance system, in which candidates can receive financial assistance from the government but which imposed rigid controls on spending. By opting out, he was able to decide his own spending strategy (Colgan 2003).

5 This excludes amorphous protests or movements which are not linked to formal, even if dispersed, organisations. These movements are increasingly significant, but are discussed on p. 116.

6 For a discussion on the use of websites for such social movements, see Tsaliki (2003).

7 The latter being an even more high profile benefit after the controversies in Florida state regarding the recounts in the Bush–Gore Presidential election of 2000.

8 The participants may be unrepresentative due to self-selection and are still only a small percentage of the total population of over three and a half million people, but a sample size of ten thousand is still likely to have predictive value. For instance, in April 2002, 72 per cent of the 8,430 participants were dissatisfied with the bishops' statement on clerical child sex abuse (*The Irish Times*, 10 April 2002). In October 2003, three out of four of 16,000 participants agreed that residents should pay to have rubbish collected, which was during a high visibility protest over such charges (*Sunday Independent*, 19 October 2003).

9 It should be noted that difficult does not mean impossible. The conflict in Iraq demonstrated that many repressive regimes still both control internal information and prevent the dissemination of external information.

10 For an example of this, see Komito (1998b).

11 With new technologies, individuals can respond rapidly and organise collective action. These are often ad hoc groups which disappear as soon as they appear, but such mobilisation about a local issue is effective and their eventual disbandment does not diminish their short-term effectiveness.

Chapter 8 State policies and the information society

1 http://news.bbc.co.uk/1/hi/technology/3222664.stm

2 As reported by news releases from the Revenue Commissioners (Keena 2003).

3 Examples would include http://www.oasis.gov.ie/, http://www.gov.uk, http://www.firstgov.gov/.

4 As recently as January 2004, Phillips announced the move of its accountacy services from Ireland to Poland, citing reduced labour charges in Poland (*The Irish Times*, 15 January 2004). This assumes that other factors (such as technology infrastructure and labour expertise) are either equal in Poland or at least close enough to make the reduced labour overheads cost-effective.

5 Shared information systems (such as electronic documents, electronic mail, video conferencing and other such technologies) make decentralised administration and decision making possible, as has been demonstrated in multinational corporations. But evidence from multinational corporations also suggests limits in terms of decision making and sharing tacit knowledge; sometimes the result of decentralised administration is the centralisation of crucial decision making to locations where face-to-face communication is still possible.

6 This definition is taken from Council of Europe Resolution 94/C48/01, but would be a commonly accepted definition.

7 This is of increasing concern in places like Dublin, where rising housing costs are driving house purchasers further and further outside central Dublin. As two hour commutes each way to work become commonplace, the prospect of avoiding such time consuming journeys by teleworking (even if only for one or two days a week) becomes more attractive.

8 This trend towards linking pay for output is not restricted to telework. Monitoring output is also possible with new ICTs in the workplace – that is, how many phone calls are made, how many customers are served (Smyth 2004 provides a example for Dublin City Council staff). In the workplace, monitoring of output is cost effective with new technology, and is being used, in addition to physical presence, as a monitoring mechanism. This makes possible a move from a regime of the

same pay scales for all people in the same job description, to determining individual pay by individual output, usually in the name of productivity.

9 In a study by TUC in the United Kingdom, 44 per cent were part time and female (http://www.tuc.org.uk/work_life/tuc-35040f0.cfm).

10 Evidence suggests that recent knowledge is not enough – software engineers fresh out of university are not actually of much benefit until they acquire about two years of experience. But after two to five years they are highly employable (see also Wickham 1998a).

11 In addition, as jobs are automated or altered through technology, training conversion from old skills to new skills is needed.

12 In some cases, the motivation for such a change can come from the purchasing company. In a recent example, an Indian soybean company has sponsored the introduction of Internet access to villages so that they can obtain the product directly. This reduces the cost and improves the efficiency of their operation. It also benefits the villages, which not only get a better price for their soybeans, but can also access information (about proper fertilisers and future weather conditions, for example) that improves their own productivity (Waldman 2004).

13 This issue will be explored in greater detail in chapter 10.

Chapter 9 Individuals and social change

1 Of course the inconveniently long queue may result from job reductions imposed by the bank, and the switch to automatic teller machines and home banking may be precisely the result which banks hoped would result from longer queues.

2 In some parts of London, it was so difficult to find a parking place near home that it was easier to get shopping delivered than to try to find parking after shopping, which meant ordering the shopping online.

3 Or shopping centre, or café, depending on the cultural milieu.

4 It also means that when an individual is composing an email or text message or placing a phone call, they can no longer know where the recipient actually is. The person answering the phone call or email message can be in Dublin or Dubai. By the same token, when reading an email or answering a phone call, the recipient cannot know whether the sender is in Delhi or Dallas. Some people find it disconcerting to talk to someone without knowing where the person actually is but this is becoming the common experience of communication.

5 For instance, there has been a move from rigid contact lenses to soft lenses that can be worn longer and now a move to laser surgery that alters eyesight permanently.

6 Statistics from Commission for Communications Regulation: Irish Communications Market, Quarterly Market Updates (http://www.comreg.ie).

7 Evidence already shows that the gender imbalance previously associated with computers and the Internet has disappeared. Class barriers remain significant but are slowly shrinking (National Telecommunications and Information Administration 2002; Central Statistics Office 2003a; Haase and Pratschke 2003).

8 Much has been written on the decreasing public trust of science and the increasing public perception that they are surrounded by risk and danger (often caused by science). The central works on 'risk' have been Beck and Ritter (1992) and Giddens (1991) but see also van Loon, Joost and Beck (2000), Toumey (1996), Downey (1986), Winner (1977), and Hess (1995).

Chapter 10 Beyond the individual

1 There are many possible definitions of community, and this topic will be further discussed later in the chapter. For the moment, it is useful to focus on community as people sharing a similar set of beliefs and understandings. Needless to say, the same logic can apply to culture as well; the crucial issue is the common understandings that develop out of interaction and there tends to be greater overlap of individual understandings at local rather than national level (Mead 1934; Archer 1988).
2 See also Colley (1992) for additional descriptions of French history.
3 The issues of national identity, nationalism and culture are far too complex and highly debated to be discussed here other than in a very superficial manner. It is enough to suggest that mass media is commonly accepted as being an important element in national identity (see Anderson 1991 for a historical perspective).
4 For a West African example of controlling mass media as well as controlling the circulation of other commodities, see Ugboajah (1985).
5 However, as the next section will explore, access to foreign media products in the United States is increasing, as ethnic groups in the United States and elsewhere use new technologies to access foreign media products that are relevant to their cultural heritage.
6 Part of the industrial development strategy of the Irish government has been to develop multimedia products that enable Irish content to be marketed abroad.
7 Although some would say it has Irish derivations originally, as Irish music traditions were exported to the United States during the waves of emigration.
8 There is ongoing debate whether electronic communication is replacing face-to-face communication, or whether both are increasing, at the expense of other activities. See Wellman and Haythornthwaite (2002) for the most recent debates on this issue.
9 Most of the statistical information about the Ennis Information Age Town comes from reports commissioned by the project group and are available at http://www.ennis.ie (e.g. McQuillan 2000).

Chapter 11 Reprise

1 According to a spokesperson for Irish Internet shopping company Buy4Now, in the run up to Christmas online shopping in 2003, 52 per cent of Irish registered Buy4Now users were women (interview on *Morning Ireland*, RTÉ, 22 December 2003).
2 As already noted, there has been a long discussion within the context of 'social studies of science and technology' about causal relations between technology, society, and individuals. In the context of the Information Society debate, it is worth focusing particularly on Castells (2000; 1996), Kling (1994; 1980), and Webster (2002; 1994), Robins and Webster (1999) as well as Duff (2000).
3 This 'international breakfast' is usually a buffet, from which guests can choose fruit, orange juice, yogurts, boiled eggs, cold meat and cheese, pickled fish, croissant, bread or dry cereal. There is something for virtually every culinary tradition, but little sense of an integrated aesthetic experience!
4 For instance, the lights on houses that were previously associated with Christmas in the United States are becoming a more common feature of Christmas in Ireland.

Bibliography

Agar, Jon, Sarah Green et al. (2002). 'Cotton to computers: from industrial to information revolutions', in S. Woolgar (ed.), *Virtual Society? Technology, Cyberbole, Reality.* Oxford, Oxford University Press: 264–85.

Anderson, Benedict (1991). *Imagined Communities: Reflections on the Origin and Spread of Nationalism.* London, Verso.

—— (1998). *The Spectre of Comparisons: Nationalism, Southeast Asia and the World.* London, Verso.

Ang, Ien (1985). *Watching Dallas: Soap Opera and the Melodramatic Imagination.* London, Methuen.

—— (1996). *Living Room Wars: Rethinking Media Audiences for a Postmodern World.* London, Routledge.

Appadurai, Arjun (1996). *Modernity at Large: Cultural Dimensions of Globalization.* Minneapolis, Minn., University of Minnesota Press.

Archer, Margaret Scotford (1988). *Culture and Agency, the Place of Culture in Social Theory.* Cambridge, Cambridge University Press.

Aronowitz, Stanley and William DiFazio (1994). *The Jobless Future: Sci-Tech and the Dogma of Work.* Minneapolis, University of Minnesota Press.

Banisar, David (2002). *Freedom of Information and Access to Government Records around the World,* Privacy International. http://www.freedominfo.org/survey/survey.pdf

Bannon, Liam (1997). 'Conceptualising the information society', *Economic and Social Review* 28 (3): 301–5.

Bannon, Liam, Ursula Barry, et al., eds (1982). *Information Technology: Impact on the Way of Life.* Dublin, Tycooly.

Barker, Martin, Jane Arthurs, et al. (2001). *The Crash Controversy: Censorship Campaigns and Film Reception.* London, Wallflower.

Barnes, John A. (1969). 'Networks and political process' in J. C. Mitchell (ed.), *Social Networks in Urban Situations.* Manchester, Manchester University Press: 51–76.

Baym, Nancy K. (1995). 'The emergence of community in computer-mediated communication', in S. G. Jones (ed.), *Cybersociety: Computer-Mediated Communication and Community.* London, Sage: 138–63.

Beck, Ulrich and Mark Ritter (1992). *Risk Society : Towards a New Modernity* London, Sage.

Behaviour & Attitudes Marketing Research (2001). *Ennis Residents Survey.* Dublin, Behaviour & Attitudes Marketing Research.

Bell, Daniel (1973). *The Coming of Post Industrial Society.* Harmondsworth: Penguin.

—— (1979). 'The social framework of the information society', in M. L. Dertouzos and J. Moses (eds), *The Computer Age: Twenty-Year View.* Cambridge, Mass, MIT Press: 163–211.

Beniger, James R. (1986). *The Control Revolution: Technological and Economic Origins of the Information Society.* Cambridge, Mass: Harvard University Press.

Benjamin, Walter (1973). 'The work of art in an age of mechanical reproduction', in W. Benjamin (ed.), *Illuminations.* London, Fontana.

Bernstein, Basil (1964). 'Aspects of language and learning in the genesis of the social process', in D. Hymes (ed.), *Language in Culture and Society: A Reader in Linguistics and Anthropology.* New York, Harper & Row: 251–63.

Bessat, John (1987). 'Information Technology and the North–South Divide', in R. Finnegan, G. Salaman and K. Thompson (eds), *Information Technology: Social Issues.* Milton Keynes, Open University: 163–80.

Bijker, Wiebe E., Thomas P. Hughes, et al., eds (1987). *The Social Construction of Technological Systems: New Directions in the Sociology and History of Technology.* Cambridge, MA, MIT Press.

Birdsall, William F. (2000). 'The digital divide in the liberal state: a Canadian perspective'. *First Monday* 5(12) http://firstmonday.org/issues/issue5_12/birdsall/index.html.

Birrer, Frans A. J. (1999). 'Cyber Society and Democratic Quality', in J. Armitage and J. Roberts (eds), *Exploring Cyber Society, Volume I.* Newcastle, UK, University of Northumbria at Newcastle.

Boyd-Barrett, Oliver, Peter Braham, et al. (1987). *Media, Knowledge and Power: A Reader.* London, Croom Helm in association with the Open University.

Braverman, Harry (1974). *Labour and Monopoly Capital.* New York, Monthly Review Press.

Brown, John Seely and Paul Duguid (2000). *The Social Life of Information.* Boston, Harvard Business School Press.

Buckland, Michael Keeble (1991). *Information and Information Systems.* New York, Greenwood Press.

Burton, Paul F. (1995). 'Regulation and control of the Internet: Is it feasible? Is it necessary?' *Journal of Information Science* 21 (6): 413–28.

Butcher, Mike (2003). '"Bloggers" aim to make an impact in the real world', *The Irish Times,* 8 August.

Cairncross, Francis (1997). *The Death of Distance.* London, Orion.

Campbell, Duncan (2001). 'How the plotters slipped US net', *Guardian,* 27 September.

Carter, Dave (1997). '"Digital democracy" or "Information aristocracy"? Economic regeneration and the information economy', in B. D. Loader (ed.), *The Governance of Cyberspace: Politics, Technology and Global Restructuring.* London, Routledge: 136–52.

Castells, Manuel (1996). *The Rise of the Network Society.* Oxford, Blackwell.

—— (1997). *The Power of Identity.* Oxford, Blackwell.

—— (1998). *End of Millennium.* Oxford, Blackwell.

—— (2000). 'Materials for an exploratory theory of the network society', *British Journal of Sociology* 51 (1): 5–24.

Castells, Manuel and Peter Geoffrey Hall (1994). *Technopoles of the World: The Making of Twenty-First-Century Industrial Complexes.* London and New York, Routledge.

Castells, Manuel and Pekka Himanen (2002). *The Information Society and the Welfare State: The Finnish Model.* Oxford, Oxford University Press.

Central Statistics Office (2003a). *Information Society Statistics, Ireland 2003.* Dublin, Stationery Office.

—— (2003b). *Quarterly National Household Survey Information and Communications Technology (ICT) June 2003.* Dublin, Central Statistics Office.

—— (2003c). *Quarterly National Household Survey: Module on Teleworking Third Quarter 2002.* Dublin, Central Statistics Office.

Chandrasekaran, Rajiv (1998a). '"Operating system" definition debated at trial', *Washington Post,* 9 December.

—— (1998b). 'U.S. uses software firm's dictionary of computer terms against IT', *Washington Post,* 10 December.

Chubb, Basil (1992). *The Government and Politics of Ireland.* London, Longman.

Clancy, Patrick, Sheelagh Drudy, et al., eds (1986). *Ireland: A Sociological Profile.* Dublin, Institute of Public Administration.

Clapham, Christopher (1982). 'Clientelism and the state', in C. Clapham (ed.), *Private Patronage and Public Power.* London, Frances Pinter: 1–35.

Clark, Colin (1940). *The Condition of Economic Progress.* Basingstoke, Macmillan.

Clarke, Donald (2002). 'The family that views together' *The Irish Times,* 21 December.

Coakley, John and Michael Gallagher, eds (1999). *Politics in the Republic of Ireland,* 3rd edn. London, Routledge.

Colgan, Jim (2003). 'Rivals out to imitate Dean's blogging campaign', *The Irish Times,* 21 November.

Colley, Linda (1992). *Britons: Forging the Nation, 1707–1837.* New Haven, CT, Yale University Press.

Collins, Richard (1996). 'The cultural dimension of communication technology and policy: the experience of satellite television in Europe', in W. H. Dufton (ed.), *Information and Communication Technologies – Visions and Realities.* Oxford, Oxford University Press: 233–48.

Commission of the European Communities (2003). *Electronic Communications: The Road to the Knowledge Economy.* Brussels, Commission of the European Communities.

Commission of the European Community (1994). *Europe and the Global Information Society: Recommendations to the European Council.* Brussels, Commission of the European Community.

—— (1996). *Building the European Information Society for Us All: First Reflections of the High Level Group of the High Level Group of Experts.* Brussels, Commission of the European Community.

—— (2000). *Europe: An Information Society for All, Progress Report.* Brussels, Commission of the European Communities. 130.

Corcoran, Mary P. (1993). *Irish Illegals Transients between Two Societies.* Westport, Conn., Greenwood Press.

Couvares, Francis G. (1996). *Movie Censorship and American Culture.* Washington and London, Smithsonian Institution Press.

Cowan, Ruth Schwartz (1983). *More Work for Mother: The Ironies of Household Technology from the Open Hearth to the Microwave.* New York, Basic Books.

Crook, John Hurrell (1972). 'The socio-ecology of primates', in S. L. Washburn and P. Dolhinow (eds), *Perspectives on Human Evolution 2.* New York, Holt, Rinehart & Winston: 281–347.

Cross, Michael (1999). 'E-voting in Africa', *The Guardian,* 10 June.

Crump, Barbara and Andrea McIlroy (2003). 'Why the 'don't-want-tos' won't compute: lessons from a New Zealand ICT project', *First Monday* 8 (12) http://firstmonday.org/issues/issue8_12/crump/index.html.

Crystal, David (1987). *The Cambridge Encyclopedia of Language.* Cambridge, Cambridge University Press.

Curran, James (1977). 'Capitalism and control of the press, 1800–1975', in J. Curran, M. Gurevitch and J. Wollacott (eds), *Mass Communication and Society.* London, Edward Arnold: 195–230.

Curran, James, Michael Gurevitch and J. Wollacott (eds) (1977). *Mass Communication and Society.* London, Edward Arnold.

Curran, James and Jean Seaton (1991). *Power without Responsibility: The Press and Broadcasting in Britain,* 4. London, Routledge.

Dandeker, Christopher (1990). *Surveillance, Power and Modernity: Bureaucracy and Discipline from 1700 to the Present Day.* Cambridge, Polity.

Daniels, Peter T. (2003). 'Writing systems', in M. Aronoff and J. Rees-Miller (eds), *The Handbook of Linguistics.* Oxford, Blackwell: 42–80.

De Grazia, Edward and K. Newman Roger (1982). *Banned Films: Movies, Censors and the First Amendment.* New York and London, Bowker.

Diffie, Whitfield and Martin Hellman (1976). 'New directions in cryptography.' *IEEE Transactions on Information Theory* 22 (6): 644–54.

Docter, Sharon and William H. Dutton (1998). 'The First Amendment online: Santa Monica's public electronic network', in R. Tsagarousianou, D. Tambini and C. Bryan (eds), *Cyberdemocracy: Technology, Cities and Civic Networks.* London, Routledge: 125–51.

Doogan, Kevin (2001). 'Insecurity and long-term employment', *Work, Employment and Society* 15 (3): 419–41.

Douglas, Mary (1973). *Natural Symbols: Explorations in Cosmology.* New York, Random House.

Downey, Gary (1986). 'Risk in culture: the American conflict over nuclear power', *Cultural Anthropology* 1 (4): 388–412.

Duff, Alistair (2000). *Information Society Studies.* London, Routledge.

Dutton, William H, Jay G Blumler, et al., eds. (1987). *Wired Cities: Shaping the Future of Communications.* Washington, DC Boston, MA, Washington Program, Annenberg School of Communications G. K. Hall.

Dutton, William H., ed. (1996). *Information and Communication Technologies – Visions and Realities.* Oxford, Oxford University Press.

—— (1999). *Society on the Line: Information Politics in the Digital Age.* Oxford, Oxford University Press.

Dutton, William H., Jay G. Blumer, et al. (1996). 'The politics of information and communication policy: the information superhighway', in W. H. Dutton (ed.), *Information and Communication Technologies – Visions and Realities.* Oxford, Oxford University Press: 387–405.

Eckert, Penelope and Sally McConnell-Ginet (1992). *Locating Power: Proceedings of the Second Berkeley Women and Language Conference,* ed. K. Hall, M. Bucholtz and B. Moonwomon. Berkeley, Calif., Berkeley Women and Language Group, University of California, Berkeley: 89–99.

Eisenstein, Elizabeth L. (1983). *The Printing Revolution in Early Modern Europe.* New York; Cambridge, Cambridge University Press.

Electronic Privacy Information Center and Privacy International (2002). *Privacy and Human Rights 2002: An International Survey of Privacy Laws and Developments.* Washington, D.C., Electronic Privacy Information Center and Privacy International.

Ellul, Jacques (1964). *The Technological Society.* New York, Random House.

Encyclopaedia Britannica (2003). *Encyclopaedia Britannica* [New edn]. Chicago, Ill. and London, Encyclopaedia Britannica.

European Commission (1996). *Universal Service for Telecommunications in the Perspective of a Fully Liberalised Environment – an Essential Element of the Information Society.* Brussels, European Commission.

European Commission Legal Advisory Board (1996). *Working Party on Illegal and Harmful Content on the Internet Report.* Brussels, European Commission.

Feather, John (1998). *The Information Society: A Study of Continuity and Change,* 2nd ed. London, Library Association Publishing.

Featherstone, Mike, ed. (1990). *Global Culture, Nationalism, Globalization and Modernity.* London, Sage.

—— (1995). *Undoing Culture: Globalization, Postmodernism and Identity.* London, Sage.

Featherstone, Mike, Scott Lash, et al., eds (1995). *Global Modernities.* London, Sage

FitzGerald, John (2000). 'The story of Ireland's failure – and belated success', in B. Nolan, P. J. O'Connell, C. T. Whelan and Institute of Public Administration (eds), *Bust to Boom? The Irish Experience of Growth and Inequality.* Dublin, Institute Public Administration: 27–57.

Forester, Tom (1987). *High-Tech Society: The Story of the Information Technology Revolution.* Cambridge, MA, MIT Press.

Forfas (1999). *E-Commerce Report.* Dublin, Department of Trade and Enterprise.

—— (2002). *Ebusiness: Where Are We and Where Do We Go from Here?* Dublin, Department of Trade and Enterprise.

Fox, Richard Gabriel, ed. (1990). *Nationalist Ideologies and the Production of National Cultures.* Washington DC, American Anthropological Association.

Frankenberg, Ronald (1969). *Communities in Britain: Social Life in Town and Country.* Harmondsworth, Penguin.

Frederick, Howard (1993). 'Networks and the emergence of global civil society', in L. M. Harasim (ed), *Global Networks: Computers and International Communication.* Cambridge, MA, MIT Press: 283–95.

Friedman, Jonathan (1994). *Cultural Identity and Global Process.* London, Sage.

Friis, Christian S. (1997). 'A critical evaluation of the Danish national ICT strategy.' *Economic and Social Review* 28 (3): 261–76.

Fuller, Steve (1997). *Science.* Buckingham, Open University Press.

Galvin, Michael (1994). 'Victory in the gulf: technology, communications and war', in L. Green and R. Guinery (eds), *Framing Technology: Society, Choice and Change.* St Leonards, NSW, Australia, Allen & Unwin: 176–90.

Garreau, Joel (1993). 'Thanksgiving in cyberspace: a far-flung, close-knit family's computer network', *Washington Post,* 25 November.

Gellner, Ernest (1983). *Nations and Nationalism.* Oxford, Blackwell.

Gibbons, Michael, Camille Limoges, et al. (1994). *The New Production of Knowlege: The Dynamics of Science and Research in Contemporary Societies.* London, Sage Publications.

Giddens, Anthony (1985). *The Nation-State and Violence.* London, Polity.

—— (1991). *The Consequences of Modernity.* Cambridge, Polity.

Gillespie, Marie (1989). 'Technology and tradition: audiovisual culture among South Asian families in west London', *Cultural Studies* 3 (2): 226–39.

Gittings, John (2001). 'In the Chinese doghouse', *The Guardian,* 27 September.

Gluckman, Max (1971). *Politics, Law and Ritual in Tribal Society.* Oxford, Basil Blackwell.

Goody, Jack (1977). *The Domestication of the Savage Mind.* Cambridge and New York, Cambridge University Press.

Gore, Al (1991). 'Infrastructure for the global village', *Scientific American* 265: 108–11.

Government of Ireland (1997). *Freedom of Information Act, 1997.* Dublin, Stationery Office.

—— (1998). *Illegal and Harmful Use of the Internet.* Dublin, Stationery Office.

—— (2002a). *Annual Report of the Information Commissioner 2001.* Dublin, Government Publications.

—— (2002b). *National Spatial Strategy for Ireland 2002–2020: People, Places and Potential.* Dublin, Stationery Office.

Graham, Stephen and Simon Marvin (1996). *Telecommunications and the City: Electronic Spaces, Urban Places.* London, Routledge.

Granovetter, Mark (1973). 'The strength of weak ties', *American Journal of Sociology* 78: 1360–80.

—— (1982). 'The strength of weak ties: a network theory revisited', in P. Marsden and N. Lin (eds), *Social Structure and Network Analysis.* Beverly Hills, CA, Sage: 105–30.

Greenfield, Patricia Marks (1984). *Mind and Media: The Effects of Television, Computers and Video Games.* London, Fontana.

Grimes, Seamus (2000). 'Rural areas in the information society: diminishing distance or increasing learning capacity?' *Journal of Rural Studies* 16 (1): 13–21.

Gross, Peter (1996). *Mass Media in Revolution and National Development: The Romanian.* Ames, Iowa, Iowa State University Press.

Grossman, Wendy (2002). 'A new blow to our privacy', *The Guardian*, 6 June.

Gurstein, Michael (2003). 'A community informatics strategy beyond the digital divide', *First Monday* 8 (12) http://firstmonday.org/issues/issue8_12/gurstein/index.html.

Haase, Trutz and Jonathan Pratschke (2003). *Digital Divide: Analysis of the Uptake of Information Technology in the Dublin Region.* Dublin, Dublin Employment Pact.

Habermas, Jurgen (1989). *The Structural Transformation of the Public Sphere: An Inquiry into a Category of Bourgeois Society.* Cambridge, MIT Press.

Hafner, Katie and Matthew Lyon (1996). *Where Wizards Stay up Late: The Origins of the Internet.* New York, Simon & Schuster.

Hammersley, Ben (2003). 'Net booms in Kabul', *The Guardian*, 16 October.

Hampton, Keith and Barry Wellman (2000). 'Examining community in the digital neighborhood: early results from Canada's wired suburb', in T. Ishida and K. K. Isbister (eds), *Digital Cities: Technologies, Experiences and Future Prospects.* Berlin, Springer Verlag: 194–208.

Hannerz, Ulf (1992). *Cultural Complexity: Studies in the Social Organization of Meaning.* New York, Columbia University Press.

—— (1996). *Transnational Connections: Culture, People, Places.* London and New York, Routledge.

Harrison, Bernice (2001). 'Tesco to sell Clubcard data', *The Irish Times*, 8 February.

Harvey, David (1989). *The Condition of Postmodernity.* Oxford, Basil Blackwell.

Haythornthwaite, Caroline and Barry Wellman (2001). 'The Internet in everyday life', *American Behavioral Scientist* 45 (3).

Herz, Jessie Cameron (1997). *Joystick Nation: How Videogames Gobbled Our Money, Won Our Hearts and Rewired Our Minds.* London, Abacus.

Hess, David J. (1995). *Science and Technology in a Multicultural World.* New York, Columbia University Press.

Hinde, Robert A. (1974). *Biological Bases of Human Social Behaviour.* New York, McGraw-Hill.

—— (1982). *Ethology: Its Nature and Relations with Other Sciences*, Fontana.

Hobsbawm, Eric and Terence Ranger, eds (1983). *The Invention of Tradition.* Cambridge, Cambridge University Press.

Hockett, Charles F. (1960). 'The origin of speech', *Scientific American*: 3–10.

Hudson, Grover (2000). *Essential Introductory Linguistics.* Oxford, Blackwell.

Huysman, Marleen, Etienne Wenger, et al., eds (2003). *Communities and Technologies.* Dordrecht and Boston, Kluwer Academic Publishers.

Information Infrastructure Task Force (1993). *National Information Infrastructure: Agenda for Action.* Washington, DC, National Telecommunications and Information Administration.

Information Society Commission (2002). *New Connections.* Dublin, Information Society Commission.

—— (2003). *Egovernment: More Than an Automation of Services.* Dublin, Information Society Commission.

Information Society Ireland (1999). *Implementing the Information Society in Ireland: An Action Plan.* Dublin, Information Society Ireland.

Information Society Steering Committee (1996). *Information Society Ireland, Strategy for Action, Report of Ireland's Information Society Steering Committee.* Dublin, Forfas.

Ingold, Tim (1987). *Evolution and Social Life*, Cambridge, Cambridge University Press.

—— (1988). 'Tools, minds and machines: an excursion into the philosophy of technology.' *Techniques et Cultures* 12: 151–76.

—— (1993). 'Tool-use, sociality and intelligence', in E. K. Gibson and T. Ingold (eds), *Tools, Language and Cognition in Human Evolution*.. Cambridge, Cambridge University Press: 429–46.

Ingold, Tim, David Riches, et al., eds (1988). *Hunters and Gatherers 1: History, Evolution and Social Change*. Oxford, Berg.

International Telecommunication Union and M. Minges (2003). ITU Digital Access Index. Geneva, International Telecommunication Union.

Jolly, Alison (1972). *The Evolution of Primate Behavior*. New York, Macmillan.

Kalathil, Shanthi and Taylor C. Boas (2001). 'The Internet and state control in authoritarian regimes: China, Cuba, and the counterrevolution.' *First Monday* 6 (8) http://firstmonday.org/issues/issue6_8/kalathil/index.html.

Katz, Raul Luciano (1988). *The Information Society, an International Perspective*. New York; London, Praeger.

Kavanaugh, Andrea L. and Scott J. Patterson (2002). 'The impact of community computer networks on social capital and community involvement in Blacksburg'. *The Internet in Everyday Life*. Oxford, UK ; Malden, MA, Blackwell Pub.: 325-44.

Keena, Colm (2003). 'Self-employed use internet to pay €200m tax in 24 hours', *The Irish Times*, 22 November.

Kelleher, Denis (2001). 'How much sense is there in censorship?' *The Irish Times*, 17 September: 14.

Kennedy, Geraldine, ed. (2002). *The Irish Times Nealon's Guide to the 29th Dail and Seanad*. Dublin, Gill & Macmillan.

Kiberd, Damien, ed. (1997). *Media in Ireland: The Search for Diversity*. Dublin, Open Air.

Kiesler, Sara, Jane Siegel, et al. (1984). 'Social psychological aspects of computer-mediated information.' *American Psychologist* 39 (10): 1123–34.

Kling, Rob (1980). 'Social analyses of computing: theoretical orientations in recent empirical research', *Computing Surveys* 12 (1): 61–110.

—— (1994). 'Reading all about computerization: how genre conventions shape nonfiction social analysis.' *The Information Society* 10: 147–72.

Koch, Howard (1970). *The Panic Broadcast: Portrait of an Event*. Boston, Little, Brown.

Kollock, Peter (1999). 'The economies of online cooperation: gifts and public goods in cyberspace', in M. A. Smith and P. Kollock (eds), *Communities in Cyberspace*. London, Routledge: 220–39.

Komito, Lee (1984). 'Irish clientelism: a reappraisal', *Economic and Social Review* 15 (3): 173–94.

—— (1989). 'Dublin politics: symbolic dimensions of clientelism', in C. Curtin and T. Wilson (eds), *Ireland from Below: Social Change and Local Communities*.. Galway, Galway University Press: 240–59.

—— (1992). 'Information technology and regional developments: promises and prospects', in J. Feehan (ed.), *Environment and Development in Ireland*. Dublin, Environmental Institute, UCD: 166–71.

—— (1997). 'Politics and administrative practice in the Irish information society', *Economic and Social Review* 28 (3): 295–300.

—— (1998a). 'The Net as a foraging society: flexible communities', *The Information Society* 14 (2): 97–106.

—— (1998b). 'Paper "work" and electronic files: defending professional practice', *Journal of Information Technology* 13: 235–46.

—— (1999). 'Political transformations: clientelism and technological change', J. Armitage and J. Roberts (eds), *Exploring Cyber Society*.. Newcastle, UK, University of Northumbria at Newcastle.

—— (2001). 'Electronic community in the information society: paradise, mirage or malaise?' *Journal of Documentation* 57 (1): 115–29.

Komito, Lee and Michael Gallagher (1999). Dáil deputies and their work', in J. Coakley and M. Gallagher (eds), *Politics in the Republic of Ireland*, 3rd edn. London, Routledge: 206–31.

Kopytoff, Igor (1986). 'The cultural biography of things: commoditization as process', in A. Appadurai (ed.), *The Social Life of Things*. Cambridge, Cambridge University Press: 64–91.

Lancaster, Jane Beckham (1975). *Primate Behavior and the Emergence of Human Culture*. New York, Holt, Rinehart & Winston.

Latour, Bruno (1987). *Science in Action: How to Follow Scientists and Engineers through Society*. Cambridge, MA Harvard University Press.

Lave, Jean and Etienne Wenger (1991). *Situated Learning: Legitimate Peripheral Participation*. Cambridge, Cambridge University Press.

Lea, Martin, ed. (1992). *Contexts of Computer-Mediated Communication*. Hemel Hempstead, Harvester Wheatsheaf.

Lee, Richard B. and Irven DeVore, eds (1968). *Man the Hunter*. Chicago, Aldine.

Lenhart, Amanda, John Horrigan, et al. (2003). *The Ever–Shifting Internet Population: A New Look at Internet Access and the Digital Divide*, The Pew Internet & American Life Project.

Levinson, Paul (1997). *The Soft Edge*. London, Routledge.

Lewellen, Ted C. (1992). *Political Anthropology: An Introduction*, 2nd ed. Westport, Conn., Bergin & Garvey.

Liebes, Tamar and Elihu Katz (1990). *The Export of Meaning: Cross-Cultural Readings of Dallas*. New York; Oxford, Oxford University Press.

Liebowitz, S. J. (2002). *Re-Thinking the Network Economy: The True Forces That Drive the Digital Marketplace*. New York, Amacom.

Lillington, Karlin (2000).' Bandwidth boost means the Republic could be key Internet player', *The Irish Times*, 25 February.

—— (2001a). 'EU battle lines drawn as security challenge to data privacy grow', *The Irish Times*, 1 June.

—— (2001b). 'Retention of mobile call records queried', *The Irish Times*, 7 November.

—— (2003). 'Our past is not so far behind us', *The Irish Times*, 23 May.

Loader, Brian D., ed. (1998). *Cyberspace Divide: Equality, Agency and Policy in the Information Society*. London, Routledge.

Lynch, Michael F. (1974). *Computer-Based Information Services in Science and Technology – Principles and Techniques*. Stevenage, Peter Peregrinus.

Lyon, David (1994). *The Electronic Eye: The Rise of Surveillance Society*. Minneapolis, University of Minnesota Press.

—— (2002). *Surveillance Society: Monitoring Everyday Life*. Buckingham and Philadelphia, Open University Press.

Lyon, David and Elia Zureik (1996). *Computers, Surveillance, and Privacy*. Minneapolis, University of Minnesota Press.

Machlup, Fritz (1962). *The Production and Distribution of Knowledge in the United States*. Princeton, N.J., Princeton University Press.

—— (1980). *Knowledge and Knowledge Production*. Princeton, N.J., Princeton University Press.

Machlup, Fritz and Kenneth Leeson (1978). *Information through the Printed Word: The Dissemination of Scholarly, Scientific, and Intellectual Knowledge*. New York, Praeger.

MacKenzie, Donald and Judy Wajcman, eds (1985). *The Social Shaping of Technology: How the Refrigerator Got Its Hum*. Milton Keynes, Open University Press.

Malik, Yasmin (2003). 'Mobile calling: mobile phones occupy a special place in the lives of Pakistanis', *The Guardian*, 21 June.

Mantovani, Giuseppe (1996). *New Communication Environments: From Everyday to Virtual.* London, Taylor & Francis.

Marvin, Carolyn (1988). *When Old Technologies Were New: Thinking About Electric Communication in the Late Nineteenth Century.* Oxford, Oxford University Press.

Masuda, Y (1981). *The Information Society as Post-Industrial Society.* Bethesda MD, World Futures Society.

Mathieson, S. A. (2002). 'Big brother gets stronger', *The Guardian*, 12 September.

—— (2003). 'X marks the spot', *The Guardian*, 1 May.

May, Christopher (2002). *The Information Society: A Sceptical View.* Cambridge, Polity.

McCaffrey, Conor (2003). *The Digital Divide in the EU: National Policies and Access to ICTs in the Member States.* Dublin, Oscail – National Distance Education Centre, Dublin City University.

McCaffrey, Una (2003). 'Educated, English-speaking and in the EU: Why Ireland's still hard to beat', *The Irish Times*, 4 July.

McDonagh, Maeve (1998). *Freedom of Information Law in Ireland.* Dublin, Round Hall Sweet & Maxwell.

McQuillan, Helen (2000). *Eircom Ennis Information Age Town: A Connected Community.* http://www.ennis.ie, Eircom Ennis Information Age Town Ltd.

Mead, George Herbert (1934). *Mind, Self, and Society: From the Standpoint of a Social Behaviorist.* Chicago, University of Chicago Press.

Melody, William H. (1996). 'Towards a framework for designing information society policies', *Telecommunications Policy* 20 (4): 243–59.

Melucci, Alberto (1996). *Challenging Codes: Collective Action in the Information Age.* Cambridge, Cambridge University Press.

Merritt, Raymond H. (1995). 'Technology', *Microsoft Encarta 96 Encyclopedia.* Richmond, Washington, Microsoft; Funk & Wagnalls.

Merton, Robert K. (1973). *The Sociology of Science: Theoretical and Empirical Investigations.* London, University of Chicago Press.

Miles, Ian, John Bessant, et al. (1987). 'IT futures in households and communities', in R. Finnegan, G. Salaman and K. Thompson (eds), *Information Technology: Social Issues..* Sevenoaks, Hodder & Stoughton, 225-42.

Mjøset, Lars (1992). *Irish Economy in Comparative Institutional Perspective.* Dublin, National Economic and Social Council.

Moore, James R. (1989). 'Communications', in C. Chant (ed.), *Science, Technology and Everyday Life: 1870–1950.* London, Routledge: 200–49.

Morgan, Lewis Henry (1877). *Ancient Society or Researches in the Lines of Human Progress from Savagery through Barbarism to Civilization.* New York, Holt, Rinehart & Winston.

Morley, David (1980). *Nationwide Audience: Structure and Decoding,* British Film Institute.

—— (1992). *Television, Audiences and Cultural Studies.* London, Routledge.

Morris, Desmond (1977). *The Naked Ape.* London, Triad Grafton.

—— (1987). *Bodywatching: A Field Guide to the Human Species.* London, Grafton.

Moseley, Fred (1999). 'The US economy at the turn of the century: entering a new era of prosperity?' *Capital and Class* 67: 25–45.

Mumford, Lewis (1934). *Technics and Civilization.* New York, Harcourt, Brace & World.

—— (1962). *Technics and Human Development: The Myth of the Machine.* New York, Harcourt Brace Jovanovich.

National Telecommunications and Information Administration (2002). *A Nation Online: How Americans Are Expanding Their Use of the Internet.* Washington, DC, United States Department of Commerce.

Navasky, Victor (1996). 'Tomorrow never knows', *New York Times*, 29 September.

Nettleton, Sarah, Nicholas Pleace, et al. (2002). 'The reality of virtual social support', in S. Woolgar (eds), *Virtual Society? Technology, Cyberbole, Reality*. Oxford, Oxford University Press: 176–88.

Nie, Norman H. (2001). 'Sociability, interpersonal relations and the Internet: reconciling conflicting findings.' *American Behavioral Scientist* 45 (3): 420–35.

Nora, Simon and Alain Minc (1980). *The Computerisation of Society*. Cambridge, MA, MIT Press.

Oakley, Kenneth Page (1957). *Man the Tool-Maker*. Chicago, University of Chicago Press.

O'Brien, Danny (2003). 'Democratic contender surfs wave of support', *The Irish Times*, 12 September.

Odasz, Frank (1995). 'Issues in the development of community cooperative networks', in B. Kahin and J. Keller (eds), *Public Access to the Internet*. Cambridge, MA, MIT Press.

O'Dea, Clare (2001). 'Shoppers give too much away with loyalty card', *The Irish Times*, 1 June.

O'Donnell, Susan, Helen McQuillan, et al. (2003). *Einclusion: Expanding the Information Society in Ireland*. Dublin, Information Society Commission.

Office of the Data Protection Commissioner (2000). *Eleventh Annual Report of the Data Protection Commissioner 1999*.

Ong, Walter J. (1982). *Orality and Literacy: The Technologizing of the Word*. London, Routledge.

Organisation for Economic Co-operation and Development (2002). *Measuring the Information Economy*. Paris, Organisation for Economic Co-operation and Development.

—— (2003a). *The E-Government Imperative*. Paris, Organisation for Economic Co-operation and Development.

—— (2003b). *The Learning Government: Introduction and Draft Results of the Survey of Knowledge Management Practices in Ministries/Departments/Agencies of Central Government*. Paris, Organisation for Economic Co-operation and Development.

—— (2003c). *OECD Communications Outlook*. Paris, Organisation for Economic Co-operation and Development.

Ó Riain, Seán (1997). 'The birth of a Celtic tiger', *Communications of the ACM* 40 (3): 1–16.

—— (1998). 'A tale of two globalizations: the Irish software industry and the global economy', *ESRI working paper 101*. Dublin, Economic and Social Research Institute.

—— (2000). 'The flexible developmental state: globalization, information technology, and the "Celtic tiger"', *Politics and Society* 28 (2): 157–93.

Ó Riain, Seán and Philip J. O'Connell (2000). 'The role of the state in growth and welfare', in B. Nolan, P. J. O'Connell, C. T. Whelan and Institute of Public Administration, *Bust to Boom? The Irish Experience of Growth and Inequality*. Dublin, Institute Public Administration: 310–39.

Orr, Julian (1996). *Talking About Machines: An Ethnography of a Modern Job*. Ithaca, NY, IRL Press.

O'Sullivan, Kevin (1999). 'Anti-GM foods campaigners undaunted', *The Irish Times*, 5 April.

Oxford English Dictionary (1998). *The New Oxford Dictionary of English*. Oxford, Oxford University Press.

Peillon, Michel (1982). *Contemporary Irish Society: An Introduction*. Dublin, Gill & Macmillan.

Peltu, Malcolm, Donald MacKenzie, et al. (1996). 'Computer power and human limits', in W. H. Dutton (ed.), *Information and Communication Technologies – Visions and Realities*. Oxford, Oxford University Press: 177–95.

PEW Internet and American Life Project (2000). Tracking Online Life: How Women Use the Internet to Cultivate Relationships with Family and Friends. http://www.pewinternet.org

—— (2001a). Online Communities: Networks That Nurture Long-Distance Relationships and Local Ties. http://www.pewinternet.org

—— (2001b). Teenage Life Online: The Rise of the Instant-Message Generation and the Internet's Impact on Friendships and Family Relationships. http://www.pewinternet.org

—— (2002). Use of the Internet at Major Life Momments. http://www.pewinternet.org

Pfaffenberger, Bryan (1992). 'Social anthropology of technology.' *Annual Review of Anthropology* 21: 491–516.

Phelps, Guy (1975). *Film Censorship*. London, Gollancz.

Pool, Ithiel de Sola (1983). *Technologies of Freedom*. Cambridge, MA, Harvard University Press.

Porat, Marc Uri, Michael Rogers Rubin, et al. (1977). *The Information Economy*. Washington, US Government Printing Office.

Poster, Mark (1990). *The Mode of Information: Poststructuralism and Social Context*. Cambridge, Polity.

Power, Carol (2001). 'Web gives more bite to word of mouth', *The Irish Times*, 22 March.

Prell, Christina (2003). 'Community Networking and Social Capital: Early Investigations', *Journal of Computer Medicated Communication* 8 (3). http://www.ascusc.org/jcmc/vol8/issue3/

Preston, Paschal (2001). *Reshaping Communications: Technology, Information and Social Change*. London, Sage.

Price, Derek J. de Solla (1963). *Little Science, Big Science*. New York, Columbia University Press.

Provenzo, Eugene F. (1991). *Video Kids: Making Sense of Nintendo*. Cambridge, MA, Harvard University Press.

Putnam, Robert D. (2000). *Bowling Alone: America's Declining Social Capital*. New York, Simon & Schuster.

Pye, Robert (1992). *An Overview of Civil Service Computerisation, 1960–1990*. Dublin, Economic and Social Research Institute.

Rheingold, Howard (1993). 'A slice of life in my virtual community', in L. M. Harasim (ed.), *Global Networks: Computers and International Communication*. Cambridge, MA, MIT Press: 57–80.

—— (1994). *The Virtual Community: Finding Connection in a Computerized World*. London, Secker & Warburg.

Roberts, Simon (1979). *Order and Dispute: An Introduction to Legal Anthropology*. Harmondsworth, Penguin.

Robertson, Roland (1992). *Globalization: Social Theory and Global Culture*. London, Sage.

Robins, Kevin and Frank Webster (1987). 'Dangers of information technology and responsibilities of education', in R. Finnegan, G. Salaman and K. Thompson (eds), *Information Technology: Social Issues*. Milton Keynes, Open University: 145–62.

—— (1999). *Times of the Technoculture: From the Information Society to the Virtual Life*. London, Routledge.

Rosenberg, Richard S. (1993). 'Free speech, pornography, sexual harassment, and electronic networks', *Information Society* 9 (4): 285–331.

Roszak, Theodore (1994). *The Cult of Information: A Neo-Luddite Treatise on High Tech, Artificial Intelligence, and the True*. Berkeley, California, University of California Press.

Scheflen, Albert (1974). *How Behaviour Means*. Garden City, NY, Anchor Press.

Schein, Louisa (2002). 'Mapping among media in diasporic space', in F. D. Ginsburg, L. Abu-Lughod and B. Larkin (eds), *Media Worlds: Anthropology on New Terrain*. Berkeley, University of California Press: 229–44.

Schiller, Herbert I. (1973). *The Mind Managers*. Boston, Beacon Press.

—— (1981). *Who Knows: Information in the Age of the Fortune 500*. Norwood, NJ, Ablex.

—— (1985). 'Electronic information flows: new basis for global domination?', in P. Drummond and D. Paterson (eds), *Television in Transition*. London: British Film Institute.

Sciadas, George (2002). *Monitoring the Digital Divide*, UNESCO, Orbicom-CIDA Project Report.

Shannon, Claude and Warren Weaver (1949). *The Mathematical Theory of Communication.* Urbana, Illinois, University of Illinois Press.

Shapiro, Andrew L. (1999). *The Control Revolution: How the Internet Is Putting Individuals in Charge and Changing the World We Know.* New York, Public Affairs.

Sheff, David (1994). *Video Games: A Guide for Savvy Parents.* New York, Random House.

Silverstone, Roger (1994). *Television and Everyday Life.* London, Routledge.

Silverstone, Roger and Eric Hirsch, eds (1992). *Consuming Technologies, Media and Information in Domestic Spaces.* London, Routledge.

Silverstone, Roger, Eric Hirsch, et al. (1991). 'Listening to a long conversation: an ethnographic approach to the study of information and communication technologies in the home.' *Cultural Studies* 5 (2): 205–27.

Slouka, Mark (1995). *War of the Worlds: Cyberspace and the Hightech Assault on Reality.* New York, Basic Books.

Smith, William John (1977). *The Behavior of Communicating: An Ethological Approach.* Cambridge, Mass, Harvard University Press.

Smyth, Jamie (2001). 'State faces decision on implementing cybercrime', *The Irish Times*, 9 November.

—— (2004). 'Dublin City Council to give its staff the 'big brother' treatment', *The Irish Times*, 23 January.

Standage, Tom (1998). *The Victorian Internet: The Remarkable Story of the Telegraph and the Nineteenth Century's on-Line Pioneers.* New York, Walker & Co.

Stehr, Nico (1994). *Knowledge Societies.* London, Sage.

Stonier, Tom (1983). *The Wealth of Information: A Profile of the Post-Industrial Economy.* London, Thames Methuen.

Suchman, Lucy (1996). 'Supporting articulation work', in R. Kling (ed.), *Computerization and Controversy: Value Conflicts and Social Choices.*. London, Academic Press: 407–23.

Surman, Mark and Kathernine Reilly (2003). *Appropriating the Internet for Social Change: Towards the Strategic Use of Networked Technologies by Transnational Civil Society Organizations.* New York, Social Science Research Council.

Tapscott, D. and A. Caston (1993). *Paradigm Shift: The New Promise of Information Technology*, New York, London, McGraw-Hill.

Taubman, G. L. (2002). 'Keeping out the Internet? non-democratic legitimacy and access to the web', *First Monday* 7 (9) http://www.firstmonday.org/issues/issue7_9/index.html.

Taylor, Cliff (2003). 'America 2003: partners making progress – US business in Ireland remains strong despite current uncertainties', *The Irish Times*, 4 July.

Taylor, Ros (1999). 'Partisans wage virtual war', *The Guardian*, 22 April.

Tinbergen, T. (1969). *The Study of Instinct.* Oxford, Oxford University Press.

Toffler, Alvin (1980). *The Third Wave.* New York, William Morrow.

Tomlinson, Bruce (1991). *Cultural Imperialism.* Baltimore, Johns Hopkins Press.

Toumey, Christopher P. (1996). *Conjuring Science: Scientific Symbols and Cultural Meanings in American Life.* New Brunswick NJ, Rutgers University Press.

Touraine, Alain (1974). *The Post-Industrial Society: Tomorrow's Social History: Classes, Conflicts and Culture in the Programmed Society.* London, Wildwood House.

Tsagarousianou, Roza, Damian Tambini, et al., eds (1998). *Cyberdemocracy: Technology, Cities and Civic Networks.* London, Routledge.

Tsaliki, Liza (2003). 'Electronic citizenship and global social movements', *First Monday* 8 (8) http://www.firstmonday.org/issues/issue8_2/.

Turkle, Sherry (1984). *The Second Self Computers and the Human Spirit.* London, Granada.

—— (1995). *Life on the Screen: Identity in the Age of the Internet.* New York, Simon & Schuster.

Ugboajah, Frank Okwu and World Association for Christian Communication (1985). *Mass Communication, Culture and Society in West Africa.* Oxford, Hans Zell.

United Nations Conference on Trade and Development (2002). *E-Commerce and Development Report 2002.* New York; Geneva, United Nations.

Van Koert, Robin (2002). 'The impact of democratic deficits on electronic media in rural development', *First Monday* 7 (4) http://www.firstmonday.org/issues/issue7_4/index.html.

van Loon, Joost, Ulrich Beck, et al., eds (2000). *The Risk Society and Beyond: Critical Issues for Social Theory.* London, Sage.

Waldman, Amy (2004). 'Indian soybean farmers join the global village', *New York Times*, 1 January 2004.

Walsham, Geoffrey (2001). *Making a World of Difference: IT in a Global Context.* Chichester and New York, J. Wiley.

Ward, Katie (2000). 'The emergence of the hybrid community: rethinking the physical virtual dichotomy', *Space and Culture* 4 (5): 71–86.

Watt, Susan E., Martin Lea, et al. (2002). 'How social is Internet communication? A reappraisal of bandwidth and anonymity effects', in S. Woolgar (ed.), *Virtual Society? Technology, Cyberbole, Reality.* Oxford, Oxford University Press: 61–77.

Weale, Sally (2002). 'The cost of reunion', *The Guardian*, 7 November.

Weber, Eugen (1977). *Peasants into Frenchmen: The Modernization of Rural France, 1870-1914.* London, Chatto & Windus.

Weber, Max (1958). *The Protestant Ethic and the Spirit of Capitalism.* New York, Scribners.

—— (1978). *Economy and Society: An Outline of Interpretive Sociology.* Berkeley, University of California Press.

Webster, Andrew (1991). *Science, Technology and Society.* London, Macmillan.

Webster, Frank (1994). 'What information society?' *The Information Society* 10: 1–23.

—— (1995). *Theories of the Information Society.* London, Routledge.

—— (2002). *Theories of the Information Society*, 2nd edn. New York, Routledge.

Webster, Frank, Raimo Blom, et al., eds (2003). *The Information Society Reader.* London, Routledge.

Webster, Juliet (1996a). 'Revolution in the office? Implications for women's paid work', in W. H. Dutton (ed.), *Information and Communication Technologies – Visions and Realities.* Oxford, Oxford University Press: 143–57.

—— (1996b). *Shaping Women's Work Gender, Employment, and Information Technology.* New York, Longman.

Weitz, Shirley, ed. (1979). *Non-Verbal Communication.* Oxford, Oxford University Press.

Wellman, Barry (1988). 'The community question re-evaluated', in M. P. Smith (ed.), *Power, Community and the City.* New Brunswick NJ, Transaction Books: 81–107.

Wellman, Barry and Caroline A. Haythornthwaite, eds (2002). *The Internet in Everyday Life.* Oxford and Malden, MA, Blackwell.

Wellman, Barry, Anabel Quan-Haase, et al. (2003). 'The social affordances of the internet for networked individualism.' *Journal of Computer Medicated Communication* 8 (3) http://www.ascusc.org/jcmc/vol8/issue3/.

Wellman, Barry, Janet Salaff, et al. (1996). 'Computer networks as social networks: collaborative work, telework, and virtual community', *Annual Review of Sociology* 22: 213–38.

Wenger, Etienne (1998). *Communities of Practice: Learning, Meaning and Identity.* Cambridge, Cambridge University Press.

Whine, Michael (1997). 'The far right on the Internet', in B. D. Loader (ed), *The Governance of Cyberspace: Politics, Technology and Global Restructuring*. London, Routledge.

Wickham, James (1987). 'Industrialisation, work and unemployment', in P. Clancy, S. Drudy, K. Lynch and L. O'Dowd (eds), *Ireland: A Sociological Profile*. Dublin, Instititute of Public Administration.

—— (1997). 'Where is Ireland in the global information society?' *Economic and Social Review* 28 (3): 277–94.

—— (1998a). 'The golden geese fly the Internet: some research issues in the migration of Irish professionals', *Economic and Social Review* 29 (1): 33–54.

—— (1998b). 'An intelligent island?' in M. Peillon and E. Slater (eds), *Encounters with Modern Ireland a Sociological Chronicle, 1995–1996*. Dublin, Institute of Public Administration: 81–8.

Wilk, Richard R. (2002). 'Television and the imaginary in Belize', in F. D. Ginsburg, L. Abu-Lughod and B. Larkin (eds), *Media Worlds: Anthropology on New Terrain*. Berkeley, University of California Press: 171–86.

Williams, Bernard Arthur Owen and Great Britain Committee on Obscenity (1979). *Report of the Committee on Obscenity and Film Censorship*. London, HMSO.

Williams, Raymond (1974). *Television, Technology and Cultural Form*. London, Fontana.

Winner, Langdon (1977). *Autonomous Technology: Technics-out-of-Control as a Theme in Political Thought*. Cambridge, MA, MIT Press.

Winston, Brian (1998). *Media Technology and Society. A History: From the Telegraph to the Internet*. London, Routledge.

Witchalls, Clint (2002). 'World wide wealth', *The Guardian*, 12 December.

Wood, Ellen Meiksins (1997). 'Modernity, postmodernity or capitalism', *Review of International Political Economy* 4 (3): 539–60.

Woodburn, James (1982). 'Egalitarian societies.' *Man* 17: 431–51.

Woolgar, Steve (1988). *Science, the Very Idea*. Chichester, London and New York, Ellis Horwood; Tavistock.

World Intellectual Property Organization (2002). *Intellectual Property on the Internet: A Survey of Issues*. World Intellectual Property Organization.

Wyatt, Sally, Graham Thomas, et al. (2002). 'They came, they surfed, they back to the beach: conceptualizing use and non-use of the Internet', in S. Woolgar (ed.), *Virtual Society? Technology, Cyberbole, Reality*. S. Woolgar. Oxford, Oxford University Press: 23–40.

Yearley, Steven (1988). *Science, Technology, and Social Change*. London, Unwin Hyman.

Young, J. Z. (1971). *An Introduction to the Study of Man*. Oxford, Oxford University Press.

Index